Abstract Domains in Constraint Programming

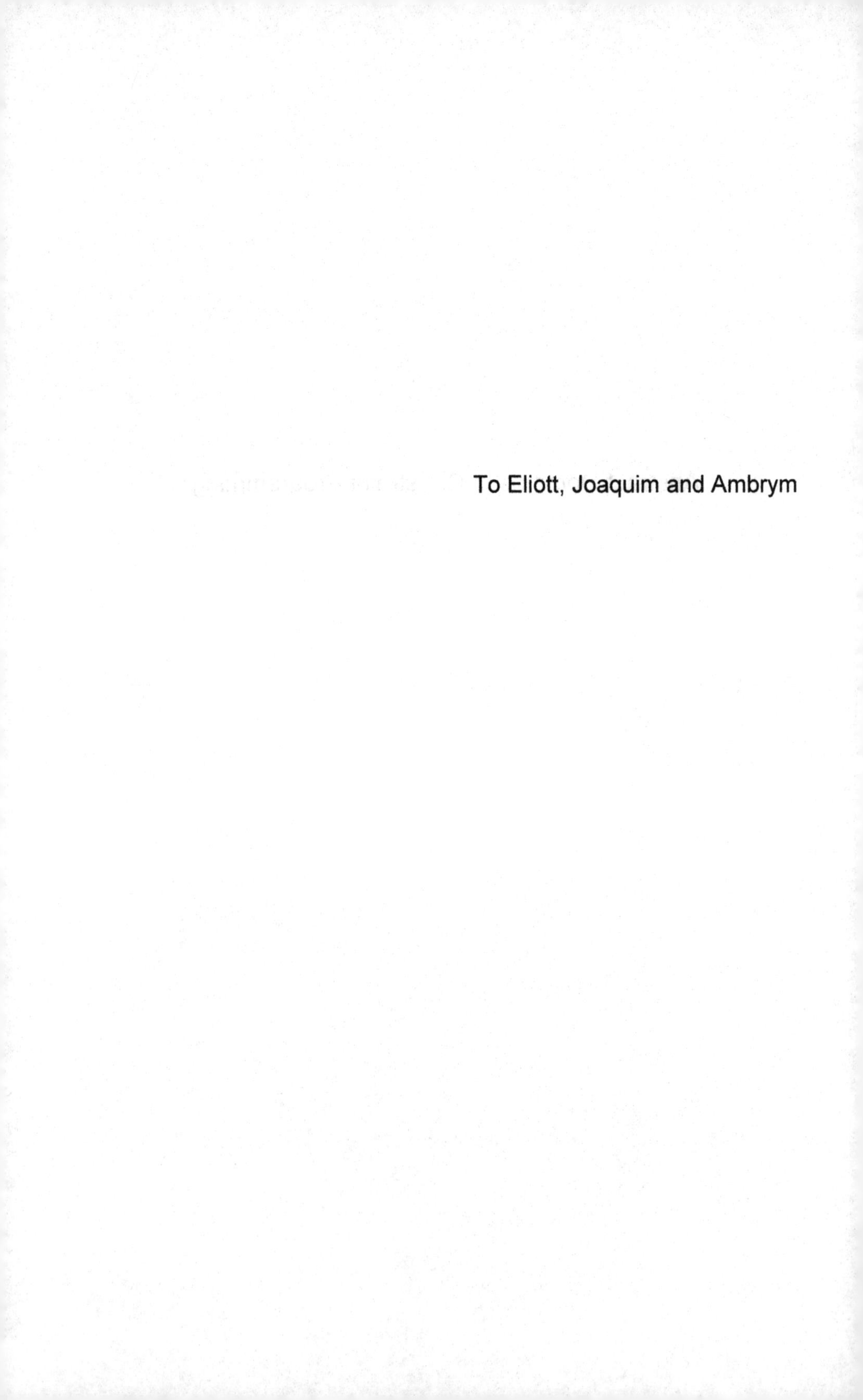

To Eliott, Joaquim and Ambrym

*Series Editor*
*Narendra Jussien*

# Abstract Domains in Constraint Programming

Marie Pelleau

First published 2015 in Great Britain and the United States by ISTE Press Ltd and Elsevier Ltd



ISTE Press Ltd
27-37 St George's Road
London SW19 4EU
UK

www.iste.co.uk

Elsevier Ltd
The Boulevard, Langford Lane
Kidlington, Oxford, OX5 1GB
UK

www.elsevier.com



**British Library Cataloguing in Publication Data**
A CIP record for this book is available from the British Library
**Library of Congress Cataloging in Publication Data**
A catalog record for this book is available from the Library of Congress
ISBN 978-1-78548-010-2

Printed and bound in the UK and US

# Contents

# Preface

Constraint programming aims at solving hard combinatorial problems, with a computation time increasing in practice exponentially. Today, the methods are efficient enough to solve large industrial problems in a generic framework. However, solvers are dedicated to a single variable type: integer or real. Solving mixed problems relies on *ad hoc* transformations. In another field, abstract interpretation offers tools to prove program properties by studying an abstraction of their concrete semantics, that is, the set of possible values of the variables during an execution. Various representations for these abstractions have been proposed. They are called abstract domains. Abstract domains can mix any type of variables, and even represent relationship between the variables. In this book, we define abstract domains for constraint programming so as to build a generic solving method, dealing with both integer and real variables. We will also study the octagons abstract domain already defined in abstract interpretation. Guiding the search by the octagonal relations, we obtain good results on a continuous benchmark. Then, we define our solving method using abstract interpretation techniques in order to include existing abstract domains. Our solver, AbSolute, is able to solve mixed problems and use relational domains.

Marie PELLEAU
February 2015

# Preface

[illegible]

# Introduction

Recent advances in computer science are undeniable. Some are visible, and others are less known to the general public: today, we are able to quickly solve many problems that are known to be difficult (requiring a long computation time). For instance, it is possible to automatically place thousands of objects of various shapes in a minimum number of containers in tens of seconds, while respecting specific constraints: accessibility of goods, non-crush, etc. [BEL 07]. Constraint programming (CP) formalizes such problems using constraints that describe a result we want to achieve (accessibility of certain objects, for example). These constraints come with efficient algorithms to solve greatly combinatorial problems. In another research area, semantics, abstract interpretation (AI) attacks an insoluble problem in the general case: the correction of programs. With strong theoretical tools developed from its creation (fixed-point theorems), AI manages to prove the properties of programs. In this area, the effectiveness of methods makes it possible for impressive applications to be solved: tools in AI have, for instance, managed to prove that there was no overflow error in the flight controls of the Airbus A380 which contains almost 500,000 lines of code.

The work presented in this book is at the interface between CP and AI, two research areas in computer science with *a priori* quite different problematics. In CP, the goal is usually to obtain a good computation time for problems that are, in general, nondeterministic polynomial time (NP), or to extend existing tools to handle more problems. In AI, the goal is to analyze very large programs by capturing a maximum of properties. Despite their differences, there is a common concern in these two disciplines: identifying an impossible or difficult (computationally) space to compute precisely (the solutions set in CP and the semantics of the program in AI). It concerns computing the relevant overapproximations of this space. CP proposes methods to carefully surround this space (consistency and propagation), always with Cartesian overapproximations (boxes in $\mathbb{R}^n$ or $\mathbb{Z}^n$). AI uses often less accurate overapproximations but not only Cartesian: they may have various different shapes (not only boxes but also octagons, ellipsoids, etc.). These non-Cartesian approximations facilitate more properties to be captured.

In this book, we exploit the similarities of these overapproximation methods to integrate AI tools in the methods of CP. We redefine tools in CP from notions of AI (abstract domains). This is not only an intellectual exercise. Indeed, by generalizing the description of overapproximations, there is a significant gain in the expressiveness of CP. In particular, the problems are treated uniformly for real and integer variables, which is not currently the case. We also develop the octagon abstract domain, showing that it is possible to exploit the relationships captured by this particular domain to solve continuous problems more effectively. Finally, we perform the opposite task: we define CP as an abstract operation in AI, and develop a solver capable of handling practically all abstract domains.

## I.1. Context

As mentioned before, the CP and AI have a common concern: computing efficiently and as accurately as possible an approximation of a difficult or impossible space. However, the issues and problems of these two areas are different, and hence so are their fields of application.

### I.1.1. *Constraint programming*

CP, whose origins date back to 1974 [MON 74], is based on the formalization of problems such as a combination of first-order logic formulas, i.e. the constraints. A constraint defines a relationship between the variables of a problem: for example, two objects placed in the same container have an empty geometric intersection, that is to say, a heavy object should be placed under a fragile object. This is known as declarative programming. CP provides efficient generic solution methods for many combinatorial problems. Academic and industrial applications are varied: job-shop scheduling problems [GRI 11, HER 11a], design of substitution tables in cryptography [RAM 11], scheduling problems [STØ 11], prediction of the ribonucleic acid (RNA) secondary structure in biology [PER 09], optical network design [PEL 09] or automatic harmonization in music [PAC 01].

One of the limitations of the expressiveness of CP methods is that they are dedicated to the nature of the problem: solvers used for discrete variable problems are fundamentally different from techniques dedicated to continuous variable problems. In a way, the semantics of the problem is different depending on whether one deals with discrete or continuous problems.

However, many industrial problems are mixed: they contain both integer and real variables. This is, for example, the case of the problem of fast power grid repair after a natural disaster [SIM 12] to restore the power as quickly as possible in the affected areas. In this problem, we try to establish a plan of action and determine the routes that should be used by repair crews. Some of the variables are discrete; for example, each device (generator, line) is associated with a Boolean variable, indicating whether it is operational or not. Others are real, as the electrical power on a line. Another example of application is the design of the topology of a multicast transmission network [CHI 08]: we want to design a network that is reliable. A network is said to be reliable when it is still effective even when one of its components is defective, so that all user communications can pass into the network with the least possible delay. Again, some of the variables are integers (the number of lines in the network) while others are continuous (the flow of information passing over the network average).

The convergence of discrete and continuous constraints in CP is both an industrial need and a scientific challenge.

### I.1.2. *Abstract interpretation*

The basis of AI was established in 1976 by Cousot and Cousot [COU 76]. AI is the theory of semantic approximation [COU 77b] in which one of the applications is programs proof. The goal is to verify and prove that a program does not contain a bug, that is to say, runtime errors. Industrial stakes are high. Indeed, many bugs have made history, such as the Year 2000 bug, or Y2K, due to system design error. On January 1, 2000, some systems showed the date of January 1, 1900. This bug may be repeated on January 19, 2038, on some UNIX systems [ROB 99]. Another example of a bug is that of the infamous inaugural flight of the Ariane 5 rocket, which, due to an error in the navigation system, caused the destruction of the rocket only 40 s after takeoff.

Every day, new softwares are being developed, corresponding to thousands or millions of lines of code. To test or verify these programs manually would require a considerable amount of time. The soundness of programs cannot be proven in a generic way; thus, AI implements methods to automatically analyze certain properties of a program. The analyzers are based on operations on the semantics of programs, that is, the set of values that can be taken by the variables of the program during its execution. By computing an overapproximation of these semantics, the analyzer can, for example, prove that the variables do not take values beyond the permitted ranges (*overflow*).

Many analyzers are developed and used for various application areas, such as aerospace [LAC 98, SOU 07], radiation [POL 06] and particle physics [COV 11].

## I.2. Problematic

In this book, we focus on CP solving methods, known as *complete*, that find the solution set or prove that it is empty, if necessary. These methods are based on an exhaustive search of the space of all possible values, also called search space. Using operations to restrict the space to visit (consistency and propagation), these methods can be accelerated. Existing methods are dedicated to a certain type of variables, discrete or continuous. Facing a mixed problem, containing both discrete and continuous variables, CP offers no real solution and the techniques available are often limited. Typically, variables are artificially transformed so that they are all discrete as in the solver Choco [CHO 10], or all continuous as in the solver RealPaver [GRA 06]. In AI, analyzed programs often, if not always, contain different types of variables. Theories of AI integrate many types of domains, and helped develop analyzers uniformly dealing with discrete and continuous variables.

We propose to draw inspiration from the work of the AI community on the different types of domains to provide new solving methods in CP. These new methods should be able, in particular, to approximate with various shapes and solve mixed problems.

## I.3. Outline of the book

This book is organized as follows: Chapter 1 gives the mandatory notions of AI and CP to understand our work and an analysis of the similarities and differences between these two research areas. Based on the similarities identified between CP and AI, we define abstract domains for CP in Chapter 2, with a resolution based on these abstract domains. The use of an example of abstract domain existing in AI in CP, the octagons, is detailed in Chapter 3. Chapter 4 deals with the solving method implementation details presented in Chapter 2 for octagons. Finally, Chapter 5 redefines the concepts of CP using the techniques and tools available in AI to define a method called abstract resolution. A prototype implementation, as well as experimental results, is finally presented.

## I.4. Contributions

The work of this book aims to design new solving techniques for CP. There are two parts in this work. In the first part, the abstract domains are defined for CP, so as mandatory operators for the solving process. These new definitions allow us to define a uniform resolution framework that no longer depends on the variables type or on the representation of the variables values. An example of a solver using the octagon abstract domain and respecting the framework is implemented in a continuous solver Ibex [CHA 09a], and tested on examples of continuous problems. In the second part, the different CP operators needed to solve are defined in AI, allowing us to define a solving method with the existing operators in AI. This method was then implemented over Apron [JEA 09], a library of abstract domains.

Most theoretical and practical results of Chapters 2–5 are the subject of publications in conferences or journals [TRU 10, PEL 11, PEL 13, PEL 14].

# 1

# State of the Art

In this chapter, we present the notions upon which abstract interpretation (AI) is based and the principles of constraint programming (CP). We do not provide an exhaustive presentation of both areas, but rather give the notions needed for the understanding of this book. The concepts discussed include those of partially ordered sets, lattice and fixpoint, which are at the basis of the underlying theories in both fields. It also includes the in-place tools, such as narrowing and widening operators in AI or consistency and splitting operators in CP. Finally, the chapter presents an analysis of the similitudes between an AI and CP upon which rely the works presented in this book.

## 1.1. Abstract interpretation

The founding principles of AI were introduced in 1976 by Patrick and Cousot [COU 76]. In this section, we only present some aspects of AI that will be needed afterward. For a more complete presentation, see [COU 92a, COU 77a].

### 1.1.1. *Introduction to abstract interpretation*

One of the applications of AI is to automatically prove that a certain type of bug does not exist in a program and that there is no error during a program execution. Let us see an example.

EXAMPLE 1.1 (Backtrace).– Consider the following program:

1: real $x, y$
2: $x \leftarrow 2$
3: $y \leftarrow 5$
4: $y \leftarrow y * (x - 2)$
5: $y \leftarrow x/y$

The backtrace for this program is:

| line | $x$ | $y$ |
|---|---|---|
| 1 | ? | ? |
| 2 | 2 | ? |
| 3 | 2 | 5 |
| 4 | 2 | 0 |
| 5 | 2 | NaN *Error: division by zero* |

In toy examples like this one, the backtrace allows us to quickly detect that the program contains errors. However, real-life programs are more complex and larger in terms of lines of code; thus, it is impossible to try all the possible executions. Moreover, the Halting's theorem states that it is undecidable to prove that a program terminates.

Nowadays, computer science is omnipresent and critical programs may contain thousands or even millions of lines of code [HAV 09]. In these programs, execution errors are directly translated into significant cost. For example, in 1996, the destruction of the Ariane 5 rocket was due to an integer overflow [LIO 96]. In 1991, American Patriot missiles failed to destroy an enemy Scud missile killing 28 soldiers due to a rounding error that had been propagated through the computations [MIS 92]. We must, therefore, ensure that such programs do not have any execution errors. Moreover, this should be done in a reasonable time without having to run the program. Indeed, sending probes into space just to check whether the program is correct, in the sense that it does not contain execution errors, is not a viable solution from an economical and ecological point of view. This is where AI comes into play. One of its applications is to verify that a program is correct during the compilation process, and thus before it is executed. The main idea is to study the values that can be taken by the variables throughout the program. We call *semantics* the set of these values and *specification* the set of all the desired behaviors such as "never divided by zero". If the semantics meets all the given specifications, then we can say that the program is correct.

An important application of AI is the design of static program analyzers that are correct and complete. An analyzer is said to be correct if it answers that a program is correct only when the program does not contain any execution error. There exist several static analyzers; we distinguish two types: the correct static analyzers, such as Astrée [BER 10] and Polyspace [POL 10], and the non-correct static analyzers, such as Coverity [COV 03]. All these analyzers have industrial applications. For instance, Astrée was able to automatically prove the absence of runtime errors in the primary flight control software of the Airbus A340 fly-by-wire system. More recently, it analyzed the electric flight control code for the Airbus A380 [ALB 05, SOU 07]. Polyspace was used to analyze the flight control code for the Ariane 502 rocket [LAC 98] and verify security softwares of nuclear installations [POL 06]. As for Coverity, it has been used to verify the code of the curiosity Mars Rover [COV 12] and ensure the accuracy of the Large Hadron Collider (LHC) software [COV 11], the particle accelerator that led to the discovery of the Higgs Boson particle.

Computing the real semantics, called *concrete semantics*, is very costly and undecidable in the general case. Indeed, Rice's theorem states that any non-trivial property formulated only on the inputs and outputs of a program is undecidable. Thus, the analyzers compute an overapproximation of the concrete semantics, the *abstract semantics*. The first step is to associate a function with each instruction of the program. This function modifies the set of possible values for the variables with respect to the instruction. Thus, the program becomes a composition of these functions and the set of observable behaviors corresponds to a fixpoint of this composition. The second step is to define an abstract domain to restrict the expressivity by keeping only a subset of properties on the program variables. An abstract domain is a computable data structure used to depict some of the program properties. Moreover, abstract domains come with efficient algorithms to compute different operations on the concrete semantics and allow the fixpoint to be computed in a finite time. The analyzer always observes a superset of program behaviors; thus, all the found properties by the analyzer are verified for the program. However, it can omit some of them.

EXAMPLE 1.2 (Deducted properties).– Consider the following program:

1: real $x, y$
2: $x \leftarrow random(1, 10)$
3: **if** $x$ mod $2 = 0$ **then**
4: $\quad y \leftarrow x/2$
5: **else**
6: $\quad y \leftarrow x - 1$
7: **end if**

As the variable $x$ takes its value between 1 and 10 (instruction 2), we can deduce that variable $y$ takes its value between 0 and 8. However, for every execution, $y < x$ is an omitted property by the intervals abstract domain but can be found with the polyhedra abstract domain.

In the following sections we present the theoretical notions upon which the AI relies.

### 1.1.2. *General presentation*

The AI underlying theory uses notions of fixpoints and lattices. These notions are recalled in this section.

#### 1.1.2.1. *Lattices*

Lattices are a well-known notion in computer science. Here, they are used to express operations on abstract domains and require some properties, such as being closed and complete.

DEFINITION 1.1 (Poset).– *A relation* $\sqsubseteq$ *on a non-empty set* $\mathcal{D}$ *is a* partial order *(Po) if and only if it is reflexive, antisymmetric and transitive. A set with a partial order is called* partially ordered set *(poset). If they exist, we denote by* $\perp$ *the least element and by* $\top$ *the greatest element of* $\mathcal{D}$.

EXAMPLE 1.3 (Partially ordered set).– Let $\mathbb{F}$ be the set of floating-point numbers according to the IEEE norm [GOL 91]. For $a, b \in \mathbb{F}$, we can define $[a, b] = \{x \in \mathbb{R}, a \leq x \leq b\}$ as the real interval delimited by the floating-point numbers $a$ and $b$, and $\mathbb{I} = \{[a, b], a, b \in \mathbb{F}\}$ as the set of intervals. Given an interval $I \in \mathbb{I}$, we denote by $\underline{I}$ (respectively, $\overline{I}$) its lower (respectively, upper) bound and for all point $x$, $\underline{x}$ (respectively, $\overline{x}$) its lower (respectively, upper) floating approximation.

Let $\mathbb{I}^n$ be the set of Cartesian products of $n$ intervals. The set $\mathbb{I}^n$ with the inclusion relation $\subseteq$ is a partially ordered set.

REMARK 1.1.– Note that the inclusion relation $\subseteq$ is a partial order. Thus, for any non-empty set $E$, $\mathcal{P}(E)$ with this relation is a partially ordered set.

DEFINITION 1.2 (Lattice).– *A partially ordered set* $(\mathcal{D}, \sqsubseteq, \sqcup, \sqcap)$ *is a* lattice *if and only if for* $a, b \in \mathcal{D}$*, the pair* $\{a, b\}$ *has a least upper bound (lub) denoted by* $a \sqcup b$*, and a greatest lower bound (glb) denoted by* $a \sqcap b$*. A lattice is said to be* complete *if and only if any subset has both a least upper bound and a greatest lower bound.*

In a lattice, any *finite* subset has a least upper bound and a greatest lower bound. In a complete lattice, any subset has a least upper bound and a greatest lower bound, even if the subset is not finite. Thus, a complete lattice has a greater element denoted by $\top$, and a least element denoted by $\perp$.

REMARK 1.2.– Notice that any finite lattice is automatically complete.

Figure 1.1 gives examples of partially ordered sets represented with Hasse diagrams. The first figure (Figure 1.1(a)) corresponds to the power sets of the set $\{1, 2, 3\}$ with the set inclusion $\subseteq$. This partially ordered set is finite and has a least element $\{\emptyset\}$ and a greatest element

$\{1,2,3\}$. Hence, it is a complete lattice. For instance, the pair $\{\{1,2\},\{1,3\}\}$ has a least upper bound $\{1,2\} \cap \{1,3\} = \{1\}$ and a greatest lower bound $\{1,2\} \cup \{1,3\} = \{1,2,3\}$.

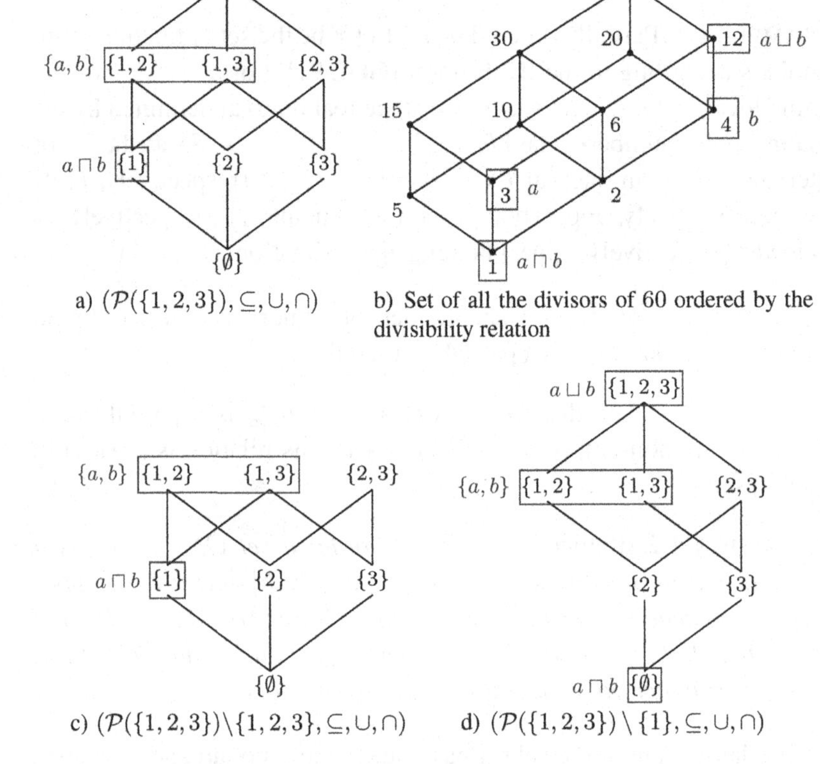


a) $(\mathcal{P}(\{1,2,3\}), \subseteq, \cup, \cap)$

b) Set of all the divisors of 60 ordered by the divisibility relation

c) $(\mathcal{P}(\{1,2,3\}) \backslash \{1,2,3\}, \subseteq, \cup, \cap)$

d) $(\mathcal{P}(\{1,2,3\}) \setminus \{1\}, \subseteq, \cup, \cap)$

**Figure 1.1.** *Examples of partially ordered sets represented with Hasse diagram*

The second figure (Figure 1.1(b)) corresponds to the set of the divisors of 60: $\{1,2,3,5,6,10,12,15,20,30,60\}$, with the divisibility relation. Similarly, this partially ordered set has a least element 1 and a greatest element 60. Thus, it is a complete lattice. The pair $\{3,4\}$ has a greatest lower bound 1 (their greatest common divisor) and a least upper bound 12 (their least common multiple).

On the contrary, the third example (Figure 1.1(c)) is not a lattice. Indeed, the pair $\{\{1,2\},\{1,3\}\}$ does not have a least upper bound. Likewise, if the element $\{\emptyset\}$ is removed from the lattice (Figure 1.1(a)), the partially ordered set obtained is no longer a lattice. However, removing any element that is neither the least nor the greatest element of the lattice does not change the fact that it is a lattice as shown in Figure 1.1(d).

EXAMPLE 1.4 (Lattice).– It is easily verified that the partially ordered set $(\mathbb{I}^n, \subseteq, \cup, \cap)$ with the least element $\bot = \emptyset$ and the greatest element $\top = \mathbb{F}^n$ is a complete lattice. Let $I = I_1 \times \cdots \times I_n$ and $I' = I'_1 \times \cdots \times I'_n$ be any two elements of $\mathbb{I}^n$. The pair $\{I, I'\}$ has a *glb*

$$I \cap I' = [\max(\underline{I_1}, \underline{I'_1}), \min(\overline{I_1}, \overline{I'_1})] \times \ldots \times [\max(\underline{I_n}, \underline{I'_n}), \min(\overline{I_n}, \overline{I'_n})]$$

and a *lub*

$$I \cup I' = [\min(\underline{I_1}, \underline{I'_1}), \max(\overline{I_1}, \overline{I'_1})] \times \ldots \times [\min(\underline{I_n}, \underline{I'_n}), \max(\overline{I_n}, \overline{I'_n})]$$

It follows that any subset has a least upper bound and a greatest lower bound, and therefore $(\mathbb{I}^n, \subseteq, \cup, \cap)$ is a lattice. Moreover, this lattice is finite; thus, $(\mathbb{I}^n, \subseteq, \cup, \cap)$ is a complete lattice.

Lattices are the base set upon which rely the abstract domains in AI. An important feature of the abstract domains is that they can be linked by a Galois connection. Note that some abstract domains do not have a Galois connection (see remark 1.5). Galois connections have been applied to the semantics by Cousot and Cousot in [COU 77a] as follows.

DEFINITION 1.3 (Galois connection).– *Let $\mathcal{D}_1$ and $\mathcal{D}_2$ be the two partially ordered sets; a Galois connection is defined by two morphisms, an abstraction $\alpha : \mathcal{D}_1 \rightarrow \mathcal{D}_2$ and a concretization $\gamma : \mathcal{D}_2 \rightarrow \mathcal{D}_1$ such that:*

$$\forall X_1 \in \mathcal{D}_1, X_2 \in \mathcal{D}_2, \alpha(X_1) \sqsubseteq X_2 \iff X_1 \sqsubseteq \gamma(X_2)$$

*Galois connections are usually represented as follows:*

$$\mathcal{D}_1 \underset{\alpha}{\overset{\gamma}{\leftrightarrows}} \mathcal{D}_2$$

REMARK 1.3.– An important consequence of this definition is that the functions $\alpha$ and $\gamma$ are monotonic for the order $\sqsubseteq$ [COU 92a], that is:

$$\forall X_1, Y_1 \in \mathcal{D}_1, X_1 \sqsubseteq Y_1 \Rightarrow \alpha(X_1) \sqsubseteq \alpha(Y_1),$$

and

$$\forall X_2, Y_2 \in \mathcal{D}_2, X_2 \sqsubseteq Y_2 \Rightarrow \gamma(X_2) \sqsubseteq \gamma(Y_2)$$

REMARK 1.4.– This definition implies that $(\alpha \circ \gamma)(X_2) \sqsubseteq X_2$ and $X_1 \sqsubseteq (\gamma \circ \alpha)(X_1)$. $X_2$ is said to be a correct approximation (or abstraction) of $X_1$.

REMARK 1.5.– Note that for a given abstract domain there can be no abstraction function. For instance, the polyhedra abstract domain has no abstraction function. Indeed, there exist an infinity of approximations of a circle with a polyhedron. Therefore, there is no Galois connection for the polyhedra abstract domain.

Figure 1.2 shows three different approximations for a circle using polyhedra. As there exists an infinity of tangent to the circle, there potentially exists a polyhedron with an infinite number of sides exactly approximating the circle.

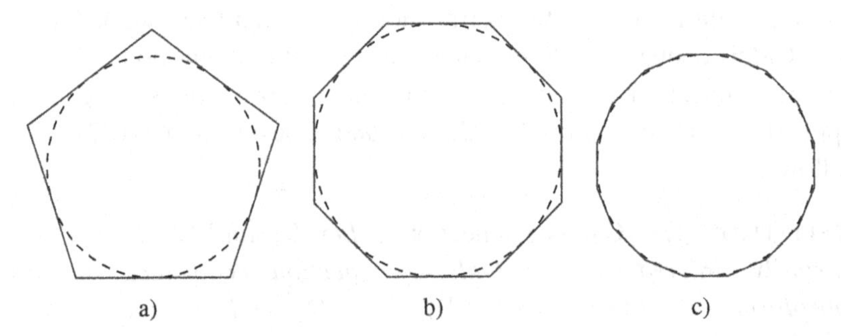


**Figure 1.2.** *Approximations for a circle*

Galois connections are used in CP even though they are not named. For instance, they are used when solving continuous problems. Indeed, as the intervals with real bounds are not computer representable, they are approximated as intervals with floating-point bounds. The transition from one representation to another forms a Galois connection as shown in the following example.

EXAMPLE 1.5 (Galois connection).– Let $\mathbb{J}$ be the set of intervals with real bounds. Given two partially ordered sets $(\mathbb{I}^n, \subset)$ and $(\mathbb{J}^n, \subset)$, there exists a Galois connection:

$$\mathbb{J}^n \underset{\alpha_{\mathbb{I}}}{\overset{\gamma_{\mathbb{I}}}{\leftrightarrows}} \mathbb{I}^n$$
$$\alpha_{\mathbb{I}}([x_1, y_1] \times \cdots \times [x_n, y_n]) = [\underline{x_1}, \overline{y_1}] \times \cdots \times [\underline{x_n}, \overline{y_n}]$$
$$\gamma_{\mathbb{I}}([a_1, b_1] \times \cdots \times [a_n, b_n]) = [a_1, b_1] \times \cdots \times [a_n, b_n]$$

In this example, the abstraction function $\alpha_{\mathbb{I}}$ transforms a Cartesian product of real bound intervals into a Cartesian product of floating-point bounds intervals. It approximates each real bound by the closest floating-point number rounded in $\mathbb{F}$ in the corresponding direction. As for the concretization function, it is straightforward since a floating-point number is also a real.

### 1.1.2.2. *Concrete/abstract*

The concrete domain, denoted by $\mathcal{D}^\flat$, corresponds to the values that can be taken by the variables throughout the program $\mathcal{D}^\flat = \mathcal{P}(V)$ with $V$ a set. Computing the concrete domain can be undecidable; thus, an approximation is accepted. The approximation of the concrete domain is called an abstract domain and is denoted by $\mathcal{D}^\sharp$. If there exists a Galois connection between the concrete domain and the abstract domain, $\mathcal{D}^\flat \underset{\alpha}{\overset{\gamma}{\leftrightarrows}} \mathcal{D}^\sharp$, then any concrete function $f^\flat$ in $\mathcal{D}^\flat$ as an abstraction $f^\sharp$ in $\mathcal{D}^\sharp$ such that

$$\forall X^\sharp \in \mathcal{D}^\sharp, (\alpha \circ f^\flat \circ \gamma)(X^\sharp) \sqsubseteq f^\sharp(X^\sharp)$$

This is a consequence of remark 1.4. Moreover, the abstract function $f^\sharp$ is said to be optimal if and only if $\alpha \circ f^\flat \circ \gamma = f^\sharp$.

In the following, we will write $\mathcal{D}$ (respectively, $f$) for a domain (respectively, function) in general (whether it is concrete or abstract).

We will write $\mathcal{D}^\flat$ and $f^\flat$ for a concrete domain and a concrete function, and $\mathcal{D}^\sharp$ and $f^\sharp$ for an abstract domain and an abstract function.

#### 1.1.2.3. *Transfer function*

In order to analyze a program, each line of code is analyzed. To do so, each instruction is associated with a function, called transfer function, which modifies the possible values for the variables.

DEFINITION 1.4 (Transfer function).– *Let $C$ be the line of code to analyze. Given an initial set of states, a transfer function $F : \mathcal{P}(\mathcal{D}) \rightarrow \mathcal{P}(\mathcal{D})$ returns a set of environments corresponding to all the possible accessible states after the execution of $C$. We will write $\{\!|C|\!\}$ the transfer function for the instruction $C$ and $\{\!|C|\!\}X$ when it is applied to the environments set $X$.*

EXAMPLE 1.6 (Transfer Function for an Affectation).– Consider the affectation $x \leftarrow expr$ with $x$ a variable and $expr$ any expression. The transfer function $\{\!|x \leftarrow expr|\!\}$ only modifies in the initial environments set the possible values for the variable $x$.

Let $x$ and $y$ be two variables taken their values in $[-10, 10]$. The transfer function $\{\!|x \leftarrow random(1, 10)|\!\}$ only modifies the values for $x$. We now have $x$ in $[1, 10]$ and $y$ in $[-10, 10]$.

EXAMPLE 1.7 (Transfer function for a condition).– Let us now consider a Boolean expression; the transfer function only keeps the environments satisfying the Boolean expression. Let $x$ and $y$ be two variables taken their values in $[-10, 10]$. For the Boolean expression $x \leq 0$, the transfer function $\{\!|x \leq 0|\!\}$ filters the values of $x$ so as to satisfy the Boolean expression. We now have $x$ in $[-10, 0]$ and $y$ in $[-10, 10]$.

REMARK 1.6.– In the following, any transfer function is supposed to be monotonic.

### 1.1.2.4. *Fixpoint*

Abstract interpretation also relies on the notion of fixpoint.

DEFINITION 1.5 (Fixpoint).– *Let $F$ be a function; we called fixpoint of $F$ an element $X$ such that $F(X) = X$. We denote by* $\mathrm{lfp}_X F$ *a least fixpoint of $F$ greater than $X$, and* $\mathrm{gfp}_X F$ *a greatest fixpoint of $F$ smaller than $X$.*

REMARK 1.7.– Note that when $F$ is monotonic, if the least or the greatest fixpoint exists, then it is unique.

Each instruction of the program is associated with a transfer function. Thus, the program corresponds to a composition of these functions. Proving that the program is correct is equivalent to computing the least fixpoint of this composition of functions. The computation of the fixpoint is mandatory to analyze loops, for instance. The functions associated with the loop are applied several times until the fixpoint is reached.

There exist several possible iterative schemas. Let $(X_1, \ldots, X_n)$ be the set of environments, where $X_i$ corresponds to the environments set for the instruction $i$. We denote by $X_i^j$ the set of environments for the instruction $i$ at the iteration $j$. Let $F_i$ be the transfer function for the instruction $i$. The most common iterative scheme is the Jacobi iteration scheme. The value of $X_i^j$ is computed using the environments sets from the previous iteration:

$$X_i^j = F_i(X_1^{j-1}, \ldots, X_n^{j-1})$$

Another iterative scheme is the Gauss–Seidel iteration scheme. It computes the value of $X_i^j$ using the environments sets already computed at the current iteration and the environments sets from the previous iteration:

$$X_i^j = F_i(X_1^j, \ldots, X_{i-1}^j, X_i^{j-1}, X_{i+1}^{j-1}, \ldots X_n^{j-1})$$

In the examples of this section, we use the Gauss–Seidel iterations.

EXAMPLE 1.8 (Gauss–Seidel iterations).– Consider the following program:

1: int $x, y$
2: $x \leftarrow 0$
3: $y \leftarrow x$
4: **while** $x < 10$ **do**
5: $\quad y \leftarrow 2x$
6: $\quad x \leftarrow x + 1$
7: **end while**

We have:

$$
\begin{aligned}
&X_1 = \top \\
&X_2 = \{\!|x \leftarrow 0|\!\} X_1 \\
&X_3 = \{\!|y \leftarrow x|\!\} X_2 \\
&X_4 = X_3 \cup X_6 \\
&X_{4'} = \{\!|x < 10|\!\} X_4 \\
&X_5 = \{\!|y \leftarrow 2x|\!\} X_{4'} \\
&X_6 = \{\!|x \leftarrow x + 1|\!\} X_5 \\
&X_7 = \{\!|x \geq 10|\!\} X_4
\end{aligned}
$$

with $X_{4'}$ the set of environments if the condition of the *while* loop is satisfied. All the environments sets are initialized at $\perp = \emptyset$, except the first environment set which is initialized to $\top = \mathbb{Z}^2$.

By applying a first time all the transfer functions, we obtain:

$$
\begin{aligned}
&X_1 = \top \\
&X_2 = \{x = 0, y \in \mathbb{Z}\} \\
&X_3 = \{x = y = 0\} \\
&X_4 = \{x = y = 0\} \\
&X_{4'} = \{x = y = 0\} \\
&X_5 = \{x = y = 0\} \\
&X_6 = \{x = 1, y = 0\} \\
&X_7 = \perp
\end{aligned}
$$

By applying a second time all the transfer functions, we obtain:

$$
\begin{aligned}
&X_1 = \top \\
&X_2 = \{x = 0, y \in \mathbb{Z}\} \\
&X_3 = \{x = y = 0\} \\
&X_4 = \{x \in [\![\mathbf{0,1}]\!], y = 0\} \\
&X_{4'} = \{x \in [\![\mathbf{0,1}]\!], y = 0\} \\
&X_5 = \{x \in [\![\mathbf{0,1}]\!], y \in [\![\mathbf{0,2}]\!]\} \\
&X_6 = \{x \in [\![\mathbf{1,2}]\!], y \in [\![\mathbf{0,2}]\!]\} \\
&X_7 = \bot
\end{aligned}
$$

where $[\![a, b]\!] = \{x \in \mathbb{Z} \,|\, a \leq x \leq b\}$ are the intervals of integers between $a$ and $b$.

The sets of environments $X_2$ and $X_3$ have not been modified and they depend on environments sets that have not been modified either. Thus, we said that their fixpoint is reached and their transfer functions are not applied anymore. On the contrary, the sets of environments $X_7$ have not been modified but depend on environments sets that have been modified ($X_4$). The other transfer functions are applied until the fixpoint is reached. After 12 iterations, the fixpoint is reached. Thus, we have:

$$
\begin{aligned}
&X_1 = \top \\
&X_2 = \{x = 0, y \in \mathbb{Z}\} \\
&X_3 = \{x = y = 0\} \\
&X_4 = \{x \in [\![0, 10]\!], y \in [\![0, 18]\!]\} \\
&X_{4'} = \{x \in [\![0, 9]\!], y \in [\![0, 18]\!]\} \\
&X_5 = \{x \in [\![0, 9]\!], y \in [\![0, 18]\!]\} \\
&X_6 = \{x \in [\![1, 10]\!], y \in [\![0, 18]\!]\} \\
&X_7 = \{x = 10, y \in [\![0, 18]\!]\}
\end{aligned}
$$

Thus, we can say that throughout this program, variable $x$ is between 0 and 10 and variable $y$ is between 0 and 18. Furthermore, after the execution of this program, the variable $x$ is equal to 10 and the variable $y$ is between 0 and 18. This result is an approximation of the possible values; indeed, at the end of the program execution, $y$ is always equal to 18.

Analyzing a program is equivalent to overapproximating the value of the environment at any point in the program. Starting from the least element $\bot$, the successive application of the transfer functions allows the fixpoint to be computed. This fixpoint corresponds to the least fixpoint of the composition of the transfer functions bigger than $\bot$. We can say that in AI, computing the least fixpoint $\text{lfp}_{\bot} F$ with $F : \mathcal{D} \to \mathcal{D}$ the composition of the transfer functions and $\mathcal{D}$ a complete partially ordered set or lattice is equivalent to analyzing the program.

However, the least fixpoint may not be reached when its computation is undecidable. In this case, an approximation is computed. Additionally, the computation of the fixpoint can converge very slowly, especially when analyzing a loop. To accelerate this process, a widening operator is used.

DEFINITION 1.6 (Widening).– *The binary abstract operator* $\nabla^\sharp$ *from* $\mathcal{D}^\sharp \times \mathcal{D}^\sharp$ *to* $\mathcal{D}^\sharp$ *is a widening if and only if:*

1) $\forall X^\sharp, Y^\sharp \in \mathcal{D}^\sharp, (X^\sharp \nabla^\sharp Y^\sharp) \sqsupseteq^\sharp X^\sharp, Y^\sharp,;$

2) *for any chain* $(X_i^\sharp)_{i\in\mathbb{N}}$*, the increasing chain* $(Y_i^\sharp)_{i\in\mathbb{N}}$ *defined by:*

$$\begin{cases} Y_0^\sharp &= X_0^\sharp \\ Y_{i+1}^\sharp &= Y_i^\sharp \nabla^\sharp X_{i+1}^\sharp \end{cases}$$

*is stable after a finite number of iterations, i.e.* $\exists K$ *such that* $Y_{K+1}^\sharp = Y_K^\sharp$.

The widening operator allows the least fixpoint lfp to be computed. It computes, as a function of several previous iterations, an approximation of the least fixpoint by staying above it in the lattice. This operator performs increasing iterations, that is, iterations that make the set of possible values for the variables grow.

REMARK 1.8.– For any partially ordered set, the widening computes an approximation of the least fixpoint in a finite number of iterations. In particular, when the considered partially ordered set has an infinite increasing chain, the widening is required to reach an approximation of the least fixpoint in a finite number of iterations. It ensures the termination of the analysis.

REMARK 1.9.– Note that if the widening and the transfer functions are correct, then the correction of the analysis is ensured.

REMARK 1.10.– If the result obtained by the widening meets the specifications, then the program is correct. In fact, if the set of environments $X$ meets the specifications, then a smaller set of environments $X' \sqsubseteq X$ also meets them. However, nothing can be deduced if the result obtained by the widening does not meet the specifications.

EXAMPLE 1.9 (Widening).– Let us consider the program in example 1.8. This time, we use a widening operator to analyze the *while* loop. Thus, we redefine the environment corresponding to the loop condition:

$$X_4 = X_4 \triangledown^\sharp (X_3 \cup X_6)$$

with $\triangledown^\sharp$ a widening operator such that:

$$[\![a, b]\!] \triangledown^\sharp [\![c, d]\!] = \left[\!\!\left[ \begin{cases} a & \text{if } a \leq c \\ -\infty & \text{otherwise} \end{cases}, \begin{cases} b & \text{if } b \geq d \\ +\infty & \text{otherwise} \end{cases} \right]\!\!\right],$$

and

$$X \triangledown^\sharp \bot = X \text{ and } \bot \triangledown^\sharp X = X$$

This operator sets the upper bound for the variable $x$ to $+\infty$ if between two iterations the interval of the variable $x$ grows toward $+\infty$. Conversely, the lower bound is set to $-\infty$ if the interval for the variable decreases.

Let us look at the evolution of the environment for $X_4$. After the first iteration, we have:

$$\begin{aligned} X_4 &= \bot \triangledown^\sharp \{x = y = 0\} \\ &= \{x = y = 0\} \end{aligned}$$

which corresponds to the result obtained after the first iteration in example 1.8. After the second iteration, we obtain:

$$\begin{aligned} X_4 &= \{x = y = 0\} \nabla^\sharp \{x \in [\![0, 1]\!], y = 0\} \\ &= \{x \in [\![0, +\infty]\!], y = 0\} \end{aligned}$$

The widening operator deduces from the previous iteration that the interval for $x$ grows in the loop and thus modifies its upper bound. Similarly, at the following iteration, the widening deduces that the interval for $y$ grows in the loop and modifies its upper bound. After two iterations, the fixpoint is reached. We now have:

$$\begin{aligned} X_1 &= \top \\ X_2 &= \{x = 0, y \in \mathbb{Z}\} \\ X_3 &= \{x = y = 0\} \\ X_4 &= \{x \in [\![0, +\infty]\!], y \in [\![0, +\infty]\!]\} \\ X_{4'} &= \{x \in [\![0, 9]\!], y \in [\![0, +\infty]\!]\} \\ X_5 &= \{x \in [\![0, 9]\!], y \in [\![0, 18]\!]\} \\ X_6 &= \{x \in [\![1, 10]\!], y \in [\![0, 18]\!]\} \\ X_7 &= \{x \in [\![10, +\infty]\!], y \in [\![0, +\infty]\!]\} \end{aligned}$$

Thus, we can say that after the execution of instruction 7, the variable $x$ is greater than or equal to 10 and the variable $y$ is positive. This is less precise than the result obtained in example 1.8. However, only four iterations were required to reach the fixpoint, which is three times less than without using the widening.

REMARK 1.11.– Note that the number of iterations required to reach the fixpoint with the widening operator does not depend on the constants in the program. By replacing the constant 10 by 100 or 1,000, the iterations without the widening will converge more slowly, while those with the widening will not be affected by this change.

This operator can generate a very large overapproximation of the fixpoint, as shown in example 1.9 where the only information that we obtained for $y$ is that it is positive, while in example 1.8, we were able to infer that $y$ was less than or equal to 18. Similarly, the variable $x$ is greater than or equal to 10, while in example 1.8, it was equal to 10. In order to refine this approximation, a narrowing operator can be used.

DEFINITION 1.7 (Narrowing).– *The abstract binary operator* $\Delta^\sharp$ *from* $\mathcal{D}^\sharp \times \mathcal{D}^\sharp$ *to* $\mathcal{D}^\sharp$ *is a narrowing if and only if:*

1) $\forall X^\sharp, Y^\sharp \in \mathcal{D}^\sharp, (X^\sharp \sqcap^\sharp Y^\sharp) \sqsubseteq^\sharp (X^\sharp \vartriangle^\sharp Y^\sharp) \sqsubseteq^\sharp X^\sharp$,;

2) *for any chain* $(X_i^\sharp)_{i\in\mathbb{N}}$, *the chain* $(Y_i^\sharp)_{i\in\mathbb{N}}$ *defined by:*

$$\begin{cases} Y_0^\sharp &= X_0^\sharp \\ Y_{i+1}^\sharp &= Y_i^\sharp \vartriangle^\sharp X_{i+1}^\sharp \end{cases}$$

*is stable after a finite number of iterations, i.e.* $\exists K$ *such that* $Y_{K+1}^\sharp = Y_K^\sharp$.

Like the widening, the narrowing deduces from the previous iterations an overapproximation of a fixpoint. However, the computed fixpoint is not necessarily the least fixpoint lfp. Indeed, the narrowing operator computes an approximation of a fixpoint (gfp or lfp) while remaining above it. Unlike the widening, the narrowing performs decreasing iterations, that is, iterations that make the set of possible values for the variables lessen.

EXAMPLE 1.10 (Narrowing).– Let us consider the program in example 1.8. In the previous example (example 1.9), we used a widening operator in order to speed up the computation of the fixpoint. Using this operator generated a large overapproximation. Starting from the result obtained in example 1.9, we use a narrowing operator to refine this result. The environment corresponding to the loop condition is thus redefined as:

$$X_4 = X_4 \vartriangle^\sharp (X_3 \cup X_6)$$

with $\vartriangle^\sharp$ a narrowing operator such that:

$$[\![a, b]\!] \vartriangle^\sharp [\![c, d]\!] = \left[\!\!\left[ \begin{cases} c \text{ if } a = -\infty \\ a \text{ otherwise} \end{cases}, \begin{cases} d \text{ if } b = +\infty \\ b \text{ otherwise} \end{cases} \right]\!\!\right],$$

and

$$X \vartriangle^\sharp \bot = \bot \text{ and } \bot \vartriangle^\sharp X = \bot$$

This operator only reduces the infinity bounds. Thus, the bounds are reduced at most one time.

Let us see the evolution of the environment $X_4$; after the first iteration, we have:

$$\begin{aligned} X_4 &= \{x \in [\![0, +\infty]\!], y \in [\![0, +\infty]\!]\} \; \Delta^\sharp \; \{x \in [\![0, 10]\!], y \in [\![0, 18]\!]\} \\ &= \{x \in [\![0, 10]\!], y \in [\![0, 18]\!]\} \end{aligned}$$

which corresponds to the result obtained after 11 iterations in example 1.8. After only two iterations, the fixpoint is reached and it is the same as in example 1.8. By using the widening and the narrowing operators, the fixpoint is reached in six iterations, which is two times less than that without using them. Note that in this example the reached fixpoint is the same in both cases. In general, if a fixpoint is reached in a finite number of iterations without using a widening and a narrowing operator, then this fixpoint may be more precise than the fixpoint obtained using these two operators.

Note that starting from example 1.9, applying the transfer functions as defined in example 1.8 generates decreasing iterations. These decreasing iterations reduce the environments without using the narrowing operator.

The schema of an approximation of a fixpoint using a widening and a narrowing operator is shown in Figure 1.3. Starting from the least element $\bot$, two increasing iterations are performed. The set of possible values for the variables grows. Then, a widening operator $\nabla^\sharp$ is used making the environment go above the least fixpoint lfp. Note that the widening may make the environment go above the greatest fixpoint gfp. The narrowing operator $\Delta^\sharp$ is then used in order to refine the obtained approximation. This decreasing iteration refines the approximation while staying above the considered fixpoint, which in our case is the least fixpoint lfp. To sum up, starting from an environment under the desired fixpoint, the widening operator makes the set of variables values grow, allowing a point above the fixpoint to be reached. As for the narrowing, given an environment above a fixpoint, it reduces the set of variables values while staying above the considered fixpoint.

While a lot of work has been devoted to designing smart widenings [BAG 05a, BAG 05b, BAG 06, SIM 06, D'SI 06, COR 08, MON 09,

SIM 10], narrowings have gathered far less attention [COU 92b, MOR 93, ALP 93, HIC 97, COR 11]. Some major domains, such as polyhedra, do not feature any. This difference of interest between the widening and the narrowing may be explained by three facts: first, narrowings are not necessary to achieve soundness unlike widenings which are mandatory. Indeed, the widening allows an approximation of the least fixpoint to be computed even if the partially ordered set has an infinite increasing chain. So, if the result obtained by the widening meets the given specifications, then the considered program is sound. However, if the result obtained with the widening does not meet the specifications, this does not mean that the program is not sound. This can be due to a too large overapproximation. Therefore, it is mandatory for the widening to be properly designed to minimize the overapproximation and thus avoid false alarms.

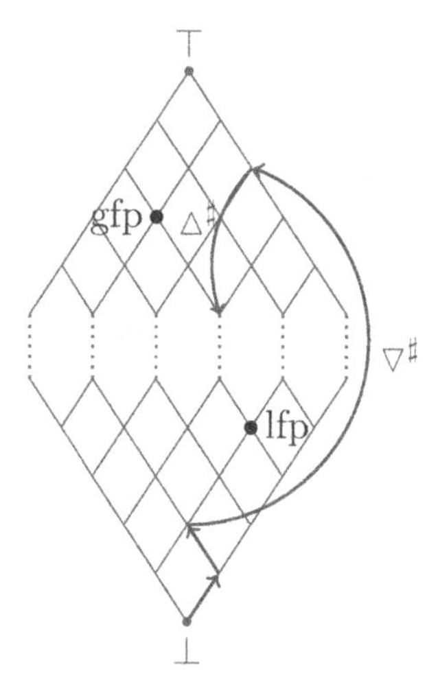


**Figure 1.3.** *Schema of an approximation of the least fixpoint using a widening and a narrowing operator*

Second, applying the transfer function without widening and narrowing operators may perform a limited number of decreasing iterations and is sometimes sufficient to recover enough precision after widening [BER 10]. This is the case for the program used to illustrate in this section (example 1.8). After the widening, the use of the transfer functions without widening reduces the environments, leading to a more precise approximation of the least fixpoint. In this example, the resulting fixpoint is the same as the one obtained with the narrowing. Moreover, the termination is guaranteed as the number of decreasing iterations is finite and generally fixed in advance.

Finally, when this simple technique is not sufficient, narrowing operators do not actually help further in practice to improve the accuracy and other methods beyond decreasing iterations must be considered such as the one presented by Halbwachs and Henry in [HAL 12]. After a series of both increasing and decreasing iterations, the fixpoint is reached; however, the achieved precision is not satisfactory. In this article, the authors propose to apply a series of iterations and keep only a portion of the previous result, in other words, to keep for each environment only the possible values for a subset of variables. The performed experiments show a reduction of the overapproximation generated by the widening operator.

Nevertheless, when the overapproximation is due to imprecise transfer functions, another solution is to use local iterations introduced by Granger in 1992 [GRA 92]. A series of decreasing iterations is applied several times. This method is detailed in the next section.

1.1.2.5. *Local iterations*

The abstract transfer function $F^\sharp \sqsupseteq (\alpha \circ F^\flat \circ \gamma)$ does not always compute efficiently the smallest abstract domain containing the considered expression, even though when it is optimal $(\gamma \circ F^\sharp \circ \alpha) = F^\flat$. To efficiently compute the result of the transfer functions, transfer functions often involve lower closure operators [GRA 92].

DEFINITION 1.8 (Lower closure operator).– *An operator $\rho : \mathcal{D} \to \mathcal{D}$ is a lower closure operator if and only if $\rho$ is:*

1) *monotonic,* $\forall X, Y \in \mathcal{D}, X \sqsubseteq Y \Rightarrow \rho(X) \sqsubseteq \rho(Y)$;

2) *reductive,* $\forall X \in \mathcal{D}, \rho(X) \sqsubseteq X$;

3) *idempotent,* $\forall X \in \mathcal{D}, \rho(X) = (\rho \circ \rho)X$.

Granger's work [GRA 92] shows that by iterating several times lower closure operators the approximation of the fixpoint may be improved. Given a correct abstraction $\rho^\sharp$ of $\rho^\flat$, the limit $Y_\delta^\sharp$ of the sequence of the narrowing operator, $Y_0^\sharp = X^\sharp$, $Y_{i+1}^\sharp = Y_i^\sharp \vartriangle \rho^\sharp(Y_i^\sharp)$ is an abstraction of $(\rho^\flat \circ \gamma)(X^\sharp)$. Note that even if $\rho^\sharp$ is not an optimal abstraction of $\rho^\flat$, $Y_\delta^\sharp$ may be significantly more precise than $\rho^\sharp(X^\sharp)$. A relevant application is the analysis of complex test conjunction $C_1 \wedge \cdots \wedge C_p$, where each atomic test $C_i$ is modeled in the abstract as $\rho_i^\sharp$. Generally, $\rho^\sharp = \rho_1^\sharp \circ \cdots \circ \rho_p^\sharp$ is not optimal, even when each $\rho_i^\sharp$ is.

Lower closure operators may be reformulated as a fixpoint. This unifies the use of narrowings and brings out the similarities in the iterative computations. Given an element $X$, $\rho$ computes the greatest fixpoint smaller than $X$, that is $\rho(X) = \mathrm{gfp}_X\, \rho$. Local iterations may be used at any time in the analysis and not only after a widening.

Given a program to analyze, an abstract domain is chosen to best represent the program properties. There exist several types of abstract domains. A brief presentation of abstract domains is given in the next section.

#### 1.1.2.6. *Abstract domains*

Abstract domains play a key role in AI. Because of the importance of numerical properties and variables in a program, many numerical abstract domains are developed. Major numerical abstract domains include intervals [COU 77a] and polyhedra [COU 78]. Recent years have seen the development of many new abstract domains capturing other properties, such as octagons [MIN 06], ellipsoids [FER 04], octahedra [CLA 04] and even varieties [ROD 04]. In addition, support libraries for abstract domains, such as Apron [JEA 09], were designed. These new domains can handle all kinds of numeric variables, mathematical integers, rationals, reals, machine integers and floating-point numbers [MIN 12]. They even express relationships between variables of different types [MIN 04], and between numeric and Boolean variables [BER 10]. Moreover, generic operators can be used to build new domains from existing abstract domains, such as disjunctive completion [COU 92a], reduced products [COU 79, COU 07, COU 11] and partitioning [BOU 92, RIV 07]. These abstract domains are the building blocks in static analyzers, which generally use different abstract domains simultaneously. It is crucial to carefully choose (or design) abstract domains for a given problem. This is usually done manually based on the properties that must be deducted.

These various representations are grouped according to their expressiveness. The more complex the expressible properties by an abstract domain are, the more it will be accurate. Expressible properties by an abstract domain correspond to the relationships between different variables that can be represented by the abstract domain. In other words, the more complex the representable relationships between variables are, the more the abstract domain is accurate. This precision usually comes at a cost in terms of computation time. There are three main categories: non-relational, relational and weakly relational domains. In the first family, the properties are expressed on a single variable, and the abstract domains of this family are the least expressive. To represent several variables, we use a Cartesian product of the selected domain. The best known in this area is the intervals abstract domain [COU 76] where each variable is represented with an interval of possible values for this variable.

In the other two families of abstract domains, we can, as their name suggests, have relationships between variables. Relational domains are very expressive. There exists a large diversity of these abstract domains. For instance, there is the polyhedra abstract domain [COU 78], which expresses linear relationships between variables, and the ellipsoids abstract domain [FER 04], which expresses second-degree polynomials of ellipses. The last family, the weakly relational domains, is a trade-off between the two families mentioned above. It is composed of abstract domains offering a trade-off between accuracy and computation time. These weakly relational domains can represent some of the possible expressible properties between variables. They were introduced in 2004 by Miné [MIN 04]. Among these abstract domains, there exists the zone abstract domain expressing inequalities of the form $v_1 - v_2 \leq c$ with $v_1$ and $v_2$ variables and $c$ a constant, and the octagon abstract domain expressing inequalities of the form $\pm v_1 \pm v_2 \leq c$.

Figure 1.4 shows for a same set of points an example of abstract domain of each of the three categories previously mentioned. The first category corresponds to a non-relational domain, the intervals (Figure 1.4(a)); the second category corresponds to a weakly relational domain, the octagons (Figure 1.4(b)); and the last category corresponds to a relational domain, the polyhedra (Figure 1.4(c)). We can see that the more complex the properties representable by an abstract domain are, the more accurate the abstract domain is. The polyhedra are more precise than the octagons which are more precise than the intervals.

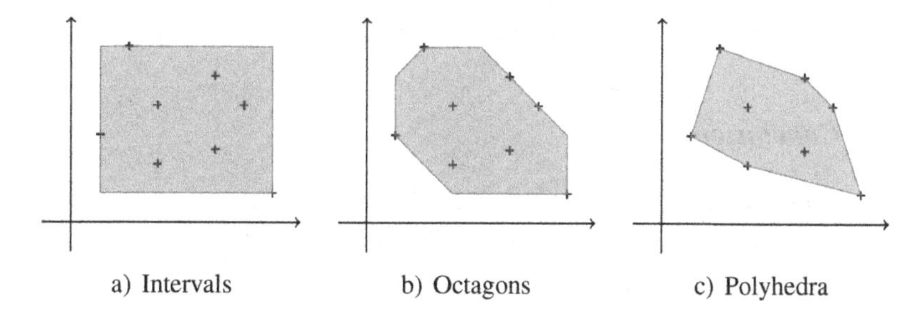


**Figure 1.4.** *Different abstract domains representing the same set of points*

An abstract domain corresponds to a computable set $\mathcal{D}^\sharp$ with the partial order $\sqsubseteq^\sharp$ to handle computable elements. Therefore, every operator in $\mathcal{D}^\flat$ must have an abstraction in $\mathcal{D}^\sharp$. These different operators are listed below.

Operators on abstract domains:

– a concretization function $\gamma : \mathcal{D}^\sharp \rightarrow \mathcal{D}^\flat$, and if it exists an abstraction function $\alpha : \mathcal{D}^\flat \rightarrow \mathcal{D}^\sharp$ forming a Galois connection $\mathcal{D}^\flat \underset{\alpha}{\overset{\gamma}{\leftrightarrows}} \mathcal{D}^\sharp$;

– a least element $\bot^\sharp$ and a greatest element $\top^\sharp$ such that $\gamma(\bot^\sharp) = \emptyset$ and $\gamma(\top^\sharp) = V$ with $\mathcal{D}^\flat = \mathcal{P}(V)$;

– efficient algorithms to compute transfer functions;

– efficient algorithms for the meet $\cup^\sharp$ and join $\cap^\sharp$;

– efficient algorithms for the widening $\nabla^\sharp$ if $\mathcal{D}^\sharp$ has an infinite increasing chain;

– if it exists and $\mathcal{D}^\sharp$ has an infinite decreasing chain, efficient algorithms for the narrowing $\Delta^\sharp$.

Note that there may be no abstraction function and no narrowing operator, and, although relatively rare, there can be no join. As mentioned in remark 1.5, the polyhedra abstract domain has no abstraction function, and therefore no Galois connection. Moreover, it does not feature any narrowing operator. Although they may not exist, these different operators are useful in order to have only one possible representation for a given concrete domain (abstraction) and to refine the overapproximation while computing the fixpoint (narrowing).

### 1.1.3. *Conclusion*

The previous sections have briefly presented AI. The main definitions and notions needed in the following are given. For a more detailed presentation, the readers are advised to refer to [COU 77a].

To sum up, AI automatically analyzes programs in order to certify that they do not contain any execution errors. This analyzing relies on the computation of the program semantic, the set of all the possible values for the variables of the program. Directly computing the concrete semantics can be very time-consuming in terms of computation time and can even be undecidable. Thus, the semantic is abstracted. In other words, only some characteristics are kept in order to simplify and speed up the process while keeping the process sound. For each concrete operator, an abstract operator is defined and efficient algorithms are designed. The abstract semantic is approximated using an abstract domain. There exist several abstract domains in AI offering different trade-offs between precision and computation time.

After this survey of AI general theory and techniques, the next section briefly presents CP, its principles, scientific challenges and resolution techniques.

## 1.2. Constraint programming

CP was introduced by Montanari [MON 74] in 1974. It relies on the idea that combinatorial problems can be expressed as a conjunction of first-order logic formulas, called constraints. A problem is called combinatorial when it contains a very large number of combinations of possible values. Take the example of a Sudoku puzzle: each cell that does not already have a value can take the value between 1 and 9. The number of all the possible grids, corresponding to the enumeration of all the possible values for all the empty boxes, is very large. If only 10 cells are empty, this number is larger than 3 million. It is equal to $9^k$, where $k$ is the number of empty cells as each empty cell has nine possible values. This is a combinatorial problem. Listing all these grids to check whether they correspond to a solution would take too much time. CP implements techniques to efficiently solve such combinatorial problems. As described by Freuder, in CP, the user specifies the problem and the computer solves it:

> *Constraint Programming represents one of the closest approaches computer science has yet made to the Holy Grail of programming: the user states the problem, the computer solves it.*
>
> – Eugene C. Freuder [FRE 97].

In order to define the problem, the user defines the constraints, which are the specifications of the problem. A constraint represents a specific combinatorial relationship of the problem [ROS 06]. With the example of Sudoku, the fact that each number between 1 and 9 appears exactly once in each row is a constraint. Each constraint comes with *ad hoc* operators using the structure expressed by the constraint to reduce combinatorics. The constraints are then combined into solving algorithms. Most research efforts in CP focus on defining and improving constraints [FAG 11], developing efficient algorithms [BES 11, PET 11] or fine tuning for the solvers [ANS 09, ARB 09]. There are many global constraints [BES 03, BEL 10] and each comes with an algorithm reducing its combinatorics.

CP provides effective techniques for combinatorial resolution with many real-life applications, among which are scheduling

[GRI 11, HER 11a], cryptography [RAM 11], rostering [STØ 11], music [TRU 11], biology [PER 09] or network design problems [PEL 09].

However, there are still limitations to the use of CP, one of the most important being the lack of solving algorithms capable of handling both discrete and continuous variables. In 2009, Berger and Granvilliers and Chabert *et al.* [BER 09, CHA 09b] proposed methods to solve mixed problems; however, the techniques used transform the variables in order to only have discrete or continuous variables.

Finally, CP offers only limited choices of variables representations which are integer or real Cartesian product of a set of bases.

In the following, we present the basics of CP necessary for the understanding of this book. We will focus only on the aspects that interest us. For a more detailed presentation, see [ROS 06].

### 1.2.1. *Principles*

In CP, the problems are modeled in a specific format, as a constraint satisfaction problem (CSP). Variables can be either integer or real.

DEFINITION 1.9 (CSP).– *A CSP is defined by a set of variables* $(v_1 \dots v_n)$ *taking values in domains* $(\hat{D}_1 \dots \hat{D}_n)$ *and a set of constraints* $(C_1 \dots C_p)$*. A constraint is a relation on a subset of variables.*

Domain $D_i$ is the set of possible values for the variable $v_i$. The set of all possible assignments for the variables $D = D_1 \times \cdots \times D_n$ is called *search space*. The search space is modified throughout the resolution; we note the initial search space $\hat{D} = \hat{D}_1 \times \cdots \times \hat{D}_n$. Problems can be either discrete ($\hat{D} \subseteq \mathbb{Z}^n$) or continuous ($\hat{D} \subseteq \mathbb{R}^n$) . Domains are always bounded in $\mathbb{R}$ or $\mathbb{Z}$.

The CSP solutions are the elements of $\hat{D}$ satisfying the constraints. We denote by $S$ the solution set $S = \{(s_1 \dots s_n) \in \hat{D} \,|\, \forall i \in$

$[\![1, p]\!], C_i(s_1 \dots s_n)\}$. For a constraint $C$, we note $S_C = \{(s_1 \dots s_n) \in \hat{D} \mid C(s_1 \dots s_n)\}$ as the solution set for $C$.

EXAMPLE 1.11 (Sudoku).– Let us consider the following $4 \times 4$ Sudoku grid:

| | | | |
|---|---|---|---|
| | 3 | 1 | |
| | | | 4 |
| 1 | 2 | | |
| | | | 1 |

A possible model is to associate to each cell a variable as follows:

| | | | |
|---|---|---|---|
| $v_1$ | $v_2$ | $v_3$ | $v_4$ |
| $v_5$ | $v_6$ | $v_7$ | $v_8$ |
| $v_9$ | $v_{10}$ | $v_{11}$ | $v_{12}$ |
| $v_{13}$ | $v_{14}$ | $v_{15}$ | $v_{16}$ |

Each variable can take a value between 1 and 4. Thus, we have $\hat{D}_1 = \hat{D}_2 = \dots = \hat{D}_{16} = [\![1, 4]\!]$. To specify that a cell has a fixed value, we can either modify its domain $\hat{D}_3 = \{1\}$ or add the constraint $v_3 = 1$. In a Sudoku grid, each row, each column and each block must contain exactly each value once. For the first row, thus we have $v_1 \neq v_2, v_1 \neq v_3, v_1 \neq v_4, v_2 \neq v_3, v_2 \neq v_4$ and $v_3 \neq v_4$. Specifying that all the variables of a subset have to be different can be done by using the global constraint alldifferent: for the first row, we can replace the six constraints listed above by the constraint $\text{alldifferent}(v_1, v_2, v_3, v_4)$.

We obtain a CSP with 16 variables $(v_1 \dots v_{16})$ taking their values in the domains $\hat{D}_2 = \{3\}, \hat{D}_3 = \hat{D}_9 = \hat{D}_{16} = \{1\}, \hat{D}_8 = \{4\}, \hat{D}_{10} = \{2\}$, all the other domains being equal to $[\![1, 4]\!]$, and 12 constraints, $(C_1 \dots C_4)$ corresponding to the rows, $(C_5 \dots C_8)$ to the columns and $(C_9 \dots C_{12})$ to the blocks:

$C_1$ : alldifferent($v_1, v_2, v_3, v_4$) $C_5$: alldifferent($v_1, v_5, v_9, v_{13}$)
$C_2$ : alldifferent($v_5, v_6, v_7, v_8$) $C_6$: alldifferent($v_2, v_6, v_{10}, v_{14}$)
$C_3$ : alldifferent($v_9, v_{10}, v_{11}, v_{12}$) $C_7$: alldifferent($v_3, v_7, v_{11}, v_{15}$)
$C_4$ : alldifferent($v_{13}, v_{14}, v_{15}, v_{16}$) $C_8$: alldifferent($v_4, v_8, v_{12}, v_{16}$)

$C_9$ : alldifferent($v_1, v_2, v_5, v_6$)
$C_{10}$: alldifferent($v_3, v_4, v_7, v_8$)
$C_{11}$: alldifferent($v_9, v_{10}, v_{13}, v_{14}$)
$C_{12}$: alldifferent($v_{11}, v_{12}, v_{15}, v_{16}$)

Solutions of this CSP correspond to all the Sudoku grids properly filled. In this example, there is one unique solution:

| | | | |
|---|---|---|---|
| 4 | 3 | 1 | 2 |
| 2 | 1 | 3 | 4 |
| 1 | 2 | 4 | 3 |
| 3 | 4 | 2 | 1 |

1.2.1.1. *Domains representation*

Depending on the variables type, the domain may be stored as a range or a set of points. In the case of an integer variable $v_i$, the domain can be represented as a subset of points $D_i$ or as a subinterval of $D_i$.

DEFINITION 1.10 (Integer Cartesian product).– *Let* $v_1, \ldots, v_n$ *be the variables with discrete finite domains* $\hat{D}_1, \ldots, \hat{D}_n$. *We call* integer Cartesian product *any Cartesian product of any integer set in* $\hat{D}$. *Integer Cartesian products of* $\hat{D}$ *form a finite lattice for the inclusion:*

$$\mathbb{S} = \left\{ \prod_i X_i \mid \forall i, X_i \subseteq \hat{D}_i \right\}$$

DEFINITION 1.11 (Integer box).– *Let* $v_1, \ldots, v_n$ *be the variables with discrete finite domains* $\hat{D}_1, \ldots, \hat{D}_n$. *We call* integer box *any Cartesian* *product of integer intervals in $\hat{D}$. Integer boxes of $\hat{D}$ form a finite lattice for the inclusion:*

$$\mathbb{IB} = \left\{ \prod_i [\![a_i, b_i]\!] \mid \forall i, [\![a_i, b_i]\!] \subseteq \hat{D}_i, a_i \leq b_i \right\} \cup \emptyset$$

For real variables, as the reals are not computer representable, their domains are represented by floating bounds intervals.

DEFINITION 1.12 (Box).– *Let $v_1, \ldots, v_n$ be the variables with bounded continuous domains $\hat{D}_1, \ldots, \hat{D}_n \in \mathbb{I}$. We call* box *any Cartesian product of floating bounds intervals in $\hat{D}$. The boxes in $\hat{D}$ form a finite lattice for the inclusion:*

$$\mathbb{B} = \left\{ \prod_i I_i \mid \forall i, I_i \in \mathbb{I}, I_i \subseteq \hat{D}_i \right\} \cup \emptyset$$

1.2.1.2. *Constraint satisfaction*

For integer variables, given an instantiation for the variables, a constraint answers *true* if this variables assignment satisfies the constraint, and *false* otherwise.

EXAMPLE 1.12 (Satisfaction of a constraint over integer variables).– Let $v_1$ and $v_2$ be two integer variables with domains $D_1 = [\![0, 2]\!]$ and $D_2 = [\![0, 2]\!]$. Let $C : v_1 + v_2 \leq 3$ be a constraint. Given the assignment of the variables $v_1$ at 2 and $v_2$ at 0, the constraint $C$ answers true, i.e. $C(2, 0)$ is true. In contrast, $C(2, 2)$ is false.

In the case of real domains, an important feature is that constraints can answer:

– *true*, if the box only contains solutions;

– *false*, if the box contains no solution at all;

– *maybe*, when we cannot determine whether the box contains solutions or not. This can happen when a box contains both solution and non-solution elements.

These different answers are due to the interval arithmetic [MOO 66]. In order to know whether a constraint is satisfied or not, each variable is replaced by its domain. Then, the constraint is evaluated, given the rules of interval arithmetic.

Let $I_1, I_2 \in \mathbb{I}$ be two intervals such that $I_1 = [a_1, b_1]$ and $I_2 = [a_2, b_2]$; we have the following formulas:

$$
\begin{aligned}
I_1 + I_2 &= \left[\, \underline{a_1 + a_2},\ \overline{b_1 + b_2}\, \right] \\
I_1 - I_2 &= \left[\, \underline{a_1 - b_2},\ \overline{b_1 - a_2}\, \right] \\
I_1 \times I_2 &= [\min\left(\underline{a_1 \times a_2}, \underline{a_1 \times b_2}, \underline{b_1 \times a_2}, \underline{b_1 \times b_2}\right), \\
&\qquad \max\left(\overline{a_1 \times a_2}, \overline{a_1 \times b_2}, \overline{b_1 \times a_2}, \overline{b_1 \times b_2}\right)] \\
I_1/I_2 &= I_1 \times [1/a_2, 1/b_2] \text{ if } 0 \notin I_2 \\
I_1^2 &= \begin{cases} \left[\min\left(\underline{a_1^2}, \underline{b_1^2}\right), \max\left(\overline{a_1^2}, \overline{b_1^2}\right)\right] & \text{if } 0 \notin I_1 \\ \left[0, \max\left(\overline{a_1^2}, \overline{b_1^2}\right)\right] & \text{otherwise} \end{cases}
\end{aligned}
$$

EXAMPLE 1.13 (Satisfaction of a constraint over real variables).– Let $v_1$ and $v_2$ be two continuous variables with domains $D_1 = [0, 2]$ and $D_2 = [0, 2]$. Consider the following three constraints:

$C_1$: $v_1 + v_2 \leq 6$;
$C_2$: $v_1 - v_2 \geq 4$;
$C_3$: $v_1 - v_2 = 0$.

Given the variables domains, the first constraint $C_1$ answers *true*. Indeed, by replacing $v_1$ and $v_2$ by their domains, we have $[0, 2] + [0, 2] = [0, 4]$, and so $[0, 4] \leq 6$ is true since any point of the interval $[0, 4]$ is less than or equal to 6. The second constraint $C_2$ answers *false*, indeed $[0, 2] - [0, 2] = [-2, 2]$ and the condition $[-2, 2] \geq 4$ is always false. As for the last constraint, it answers *maybe*: we have $[-2, 2] = 0$, the only possible deduction is that there are perhaps values of $v_1$ and $v_2$ for which the constraint is satisfied.

#### 1.2.1.3. *Solutions and approximations*

Given a problem, we try to solve it, that is, to find a solution. In the case of discrete variables, a solution is an instantiation of all variables, which corresponds to have a single value in each domain. In the case of continuous variables, a solution is considered a box containing only solutions or small enough. The fact that a solution is contained in small solution box is unclear. By being small, it reduces the chances of not containing any solution, but does not delete them.

Solving a CP problem is equivalent to computing the solution set or an approximation of it. In the case of discrete variables, listing all the solutions is possible but can be expensive. Depending on the application, when a problem has a large number of solutions, we do not necessarily want to list them all. For example, consider $x \in \mathbb{Z}$ as an integer variable, the constraints $x \geq 0, x \leq 999$ have a very large number of solutions $(1,000)$, and we can only want one or a subset of solutions. For continuous variables, the set of solutions can be infinite and to list them is impossible. For example, it is impossible to enumerate the real numbers between $0$ and $1$. Moreover, for a finite set of solutions, it is unlikely that the actual solutions are computer representable. In this case, an approximation of the solution set is accepted.

DEFINITION 1.13 (Approximation).– *A* complete approximation *(respectively,* sound approximation*) of the solution set is a union of domain sequences or products $D_1 \ldots D_n$ such that $D_i \subseteq \hat{D}_i$ and $S \subseteq \bigcup(D_1 \times \cdots \times D_n)$ (respectively, $\bigcup(D_1 \times \cdots \times D_n) \subseteq S$).*

Soundness guarantees that we find only solutions, while completeness guarantees that no solution is lost. On discrete domains, constraint solvers are expected to be sound and complete, i.e. compute the exact set of solutions. This is generally impossible on continuous domains given that the reals are not computer representable, and we usually withdraw either soundness (most of the time) or completeness. In the first case, the resolution returns boxes that may contain points that are not solutions [BEN 99]; we speak of outer approximation or *overapproximation*. This approximation is the most common. For most problems, we seek to know all the answers, and more importantly we want to be sure not to lose any.

In the second case, all the returned boxes contain only solutions [COL 99, CHR 06]. In this case, we speak of inner approximation or *underapproximation*. This type of approximation appears in problems where we want to be sure that all points are good solutions; for example, when controlling a robot arm in surgery [ABD 96], we want to ensure that the robot arm remains in the area of the operation. Another example is the control of a camera [BEN 04]. In this problem, we want to determine the movements that must be performed by a camera (travel, zoom, etc.) to achieve an animation of a scene and a plan type specified by the user.

REMARK 1.12.– The notions of correctness and completeness are different in AI and CP. To avoid ambiguity, we use the terms "overapproximation" for a CP-complete AI-sound approximation and "underapproximation" for an AI-complete CP-sound approximation.

To solve a problem, that is, to compute the set of solutions (or an approximation in the case of reals) of the CSP, the CP-complete solving methods alternate two phases: propagation and exploration.

### 1.2.2. *Propagation*

First, propagation algorithms try to reduce the domains based on the constraints. The domain values that cannot be part of a solution are deleted from domains. These values are called inconsistent values.

#### 1.2.2.1. *Consistency for one constraint*

Given a constraint $C$ and domains, the consistency deletes all the inconsistent values for the constraint $C$ from the domains. Several versions of consistency have been proposed, such as *generalized arc consistency* (GAC), also known as *hyper-arc consistency* or *domains consistency* [MAC 77b], *path-consistency* [MON 74], *k-consistency* [FRE 78, FRE 82] or even *bound consistency*, also called *interval consistency* [DEC 03, HEN 95, APT 03, SCH 05]. They differ according to the domains type and their "strength". The strength is evaluated by the number of inconsistent values deleted from the domains. The stronger the consistency is, the more expensive it is in terms of computation time and memory. We will return to this point later. These consistencies are based on the notion of support.

DEFINITION 1.14 (Support).– *Let $v_1 \dots v_n$ be the variables on finite discrete domains $D_1 \dots D_n, D_i \subseteq \hat{D}_i$, and $C$ a constraint. The value $x_i \in D_i$ has a* support *if and only if $\forall j \in [\![1, n]\!], j \neq i, \exists x_j \in D_j$ such that $C(x_1, \dots, x_n)$ is true.*

The most usual consistencies are given in the following.

DEFINITION 1.15 (GAC).– *Given variables $v_1 \dots v_n$ on finite discrete domains $D_1 \dots D_n, D_i \subseteq \hat{D}_i$, and $C$ a constraint. The domains are called GAC for $C$ if and only if $\forall i \in [\![1, n]\!], \forall x \in D_i$, $x$ has a support.*

The GAC only keeps the values for which there is a solution for the constraint $C$, namely, values having a support.

REMARK 1.13.– This consistency is also known as hyper-arc consistency or domains consistency [MAC 77b].

EXAMPLE 1.14 (GAC).– Let $(v_1, v_2)$ be two variables on discrete domains $D_1 = D_2 = \{-1, 0, 1, 2, 3, 4\}$ and $v_1 = 2v_2 + 2$ be a constraint. The arc-consistent domains for this CSP are $D_1 = \{0, 2, 4\}$ and $D_2 = \{-1, 0, 1\}$.

Values $-1$ and $1$ have been removed from $v_1$ domain because there exists no value for $v_2$ such that the constraint is satisfied. Similarly, if $v_2 \geq 2$, we can deduce that $v_1 \geq 6$. However, the maximum value for $v_1$ is 4. Thus, the values greater than or equal to 2 can be removed from the domain of $v_2$.

The GAC for a binary constraint $C$ has a time complexity in the worst case of $O(d^2)$ [MOH 86], with $d$ the size of the largest domain. Note that this complexity does not take into account the complexity of the constraint. For each variable, every value must be tested in order to ensure that they have a support. For constraints with more than two variables, verifying that the domains are arc-consistent is nondeterministic polynomial time (NP)-complete, even when the constraints are linear [CHO 06]. However, there are specific algorithms for certain types of constraints. For instance, there exists a dedicated algorithm for different versions of the alldifferent global constraint [SEL 02, HOE 04, GEN 08].

Other consistencies on discrete variables have been defined, such as the bound consistency based on the integer box representation. It is defined as follows.

DEFINITION 1.16 (Bound consistency).– *Let* $v_1 \dots v_n$ *be the variables on finite discrete domains* $D_1 \dots D_n, D_i \subseteq \hat{D}_i$, *and* $C$ *a constraint. The domains are said to be* bound-consistent *(BC) for* $C$ *if and only if* $\forall i \in [\![1, n]\!]$, $D_i$ *is an integer interval,* $D_i = [\![a_i, b_i]\!]$, *and* $a_i$ *and* $b_i$ *have a support.*

This consistency is weaker than the GAC; in fact, it only checks whether the bounds of the domains have a support or not.

REMARK 1.14.– There exist many different definitions of the bound-consistency [CHO 06] (bound(D)-consistency, bound(Z)-consistency, bound(R)-consistency or even range consistency). The definition we defined above corresponds to the bound(Z)-consistency.

EXAMPLE 1.15 (Bound consistency).– Let $(v_1, v_2)$ be two variables on discrete domains $D_1 = D_2 = [\![-1, 4]\!]$ and $v_1 = 2v_2 + 2$ be a constraint. The bound-consistent domains for this CSP are $D_1 = [\![0, 4]\!]$ and $D_2 = [\![-1, 1]\!]$.

The bound-consistent and arc-consistent domains $D_2$ for the variable $v_2$ are the same ($-1$, $0$ and $1$) although the representations are different (integer interval or set). On the contrary, the bound-consistent domain $D_1 = [\![0, 4]\!]$ for the variable $v_1$ is a more compact representation than the arc-consistent domain $D_1 = \{0, 2, 4\}$ but contains more values for $v_1$.

Intuitively, we can say that the complexity of computing the bound consistency for a binary constraint is in the worst case $O(d^2)$ with $d$ the maximum domains size. Indeed, in the worst case, the bound consistency keeps only one value in the domains. Thus, it has to check whether each value has a support. Deciding bound consistency of a constraint is NP-hard [SCH 05, CHO 06]. Note that there exist several studies giving effective algorithms to compute the bound consistency of specific constraints, such as the global constraint alldifferent [PUG 98, MEH 00, LÓP 03], the global constraint gcc [KAT 03, QUI 03] or set constraints [HEN 08].

There are also different consistencies for continuous variables domains, such as *hull-consistency* (HC), *box-consistency* or the extension of the GAC for continuous constraints [BEN 94]. We present here only the HC as defined in [BEN 97b].

DEFINITION 1.17 (HC).– *Let $v_1 \dots v_n$ be the variables on continuous domains represented with intervals $D_1 \dots D_n \in \mathbb{I}, D_i \subseteq \hat{D}_i$, and $C$ a constraint. Domains $D_1 \dots D_n$ are said to be HC for $C$ if and only if $D_1 \times \dots \times D_n$ is the smallest box with floating-point bounds containing the solutions for $C$ in $D_1 \times \dots \times D_n$.*

EXAMPLE 1.16 (HC).– Consider two variables $(v_1, v_2)$ on continuous domains $D_1 = D_2 = [-1, 4]$ and the constraint $v_1 = 2v_2 + 2$. The hull-consistent domains for this CSP are $D_1 = [0, 4]$ and $D_2 = [-1, 1]$.

There exist several algorithms computing the HC for a constraint. In 1999, Benhamou *et al.* [BEN 99] proposed an algorithm (HC4-Revise) based on the tree representation of the constraint in order to compute the HC in a linear time $O(e)$ with $e$ the number of unary or binary operators in the constraint. More recently, a new algorithm Mohc-Revise [ARA 10] studies the monotony of the constraint so as to compute the HC.

EXAMPLE 1.17 (HC4-Revise).– Let us consider the CSP from example 1.16. The constraint $v_1 = 2v_2 + 2$ is represented by the following syntax tree:

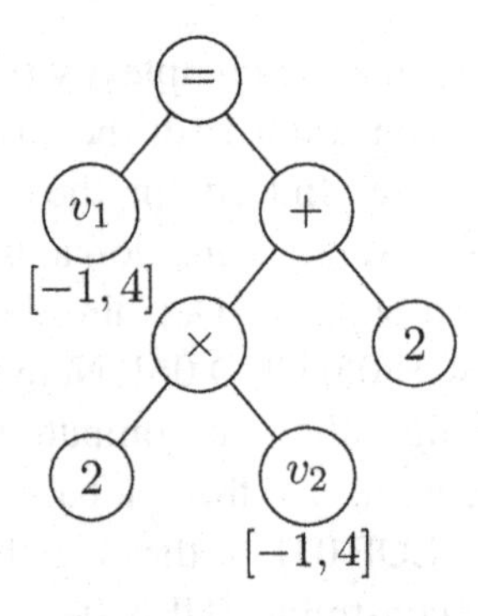


The nodes corresponding to a variable are tagged with the corresponding variable domain. The HC4-Revise algorithm performs a first pass in tree from the leaves to the root. During this first pass, it uses the interval arithmetic [MOO 66] in order to tag each node by its interval of possible values. For instance, the node $(\times)$ is tagged $2 \times [-1, 4] = [-2, 8]$. The root tag is computed using its children, given the type of constraints. In this example, the constraint being an equality, the root tag is equal to the intersection of its children nodes' interval.

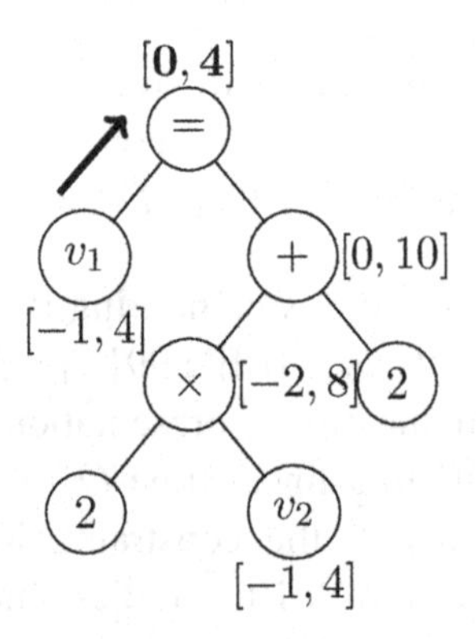


Then, a second pass is performed, but this time from the root to the leaves. It gives the information obtained at the root to all the nodes and can modify the possible values at each node. The node $(\times)$ is now tagged $[0, 4] - 2 = [-2, 2]$.

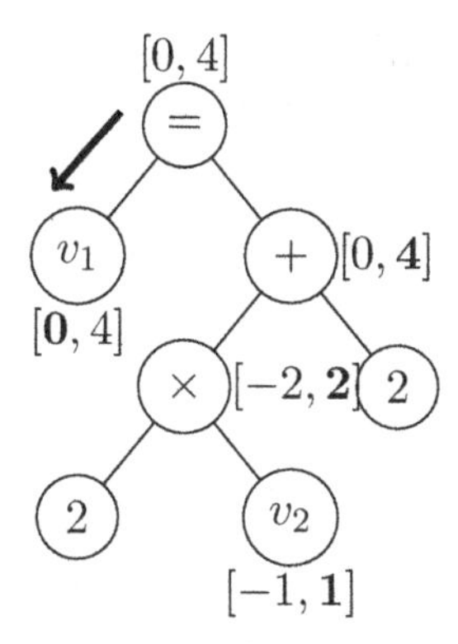


By applying this algorithm, we obtain the same results as in example 1.16 ($D_1 = [0, 4], D_2 = [-1, 1]$).

REMARK 1.15.– Note that this is not always the case, as the HC4-Revise algorithm can compute an approximation of the HC. For instance, when there exist multiple occurrences of a variable in the constraint, the result obtained with the HC4-Revise algorithm can be an overapproximation of the hull-consistent domains.

Figure 1.5 shows the different consistencies obtained in examples 1.14, 1.15 and 1.16. In Figures 1.5(a) and 1.5(b), the points correspond to the Cartesian product of the consistent domains. There are three solution points and are represented by crosses. With the Cartesian product of the arc-consistent domains (Figure 1.5(a)), we obtain 9 points, while with the bound-consistent domains (Figure 1.5(b)), we obtain 15 points. We can deduce that the stronger the chosen consistency is, the higher the ratio of number of solution points on the total number of points is. Figure 1.5(c) shows the hull-consistent box. The solutions correspond to the diagonal in the box. The solutions are also represented and correspond to the diagonal line in the box. We can see that in this box there are more non-solution points than solution points.

For a given constraint $C$, an algorithm taking the domains as input and removing from them the inconsistent values is called a *propagator* and is denoted by $\rho_C$. The HC4-Revise algorithm is a propagator for the HC. The propagators usually return an overapproximation of $S$. Sometimes, they compute exactly the consistent domains, but as this can be very expensive or intractable, they generally return an overapproximation of the consistent domains.

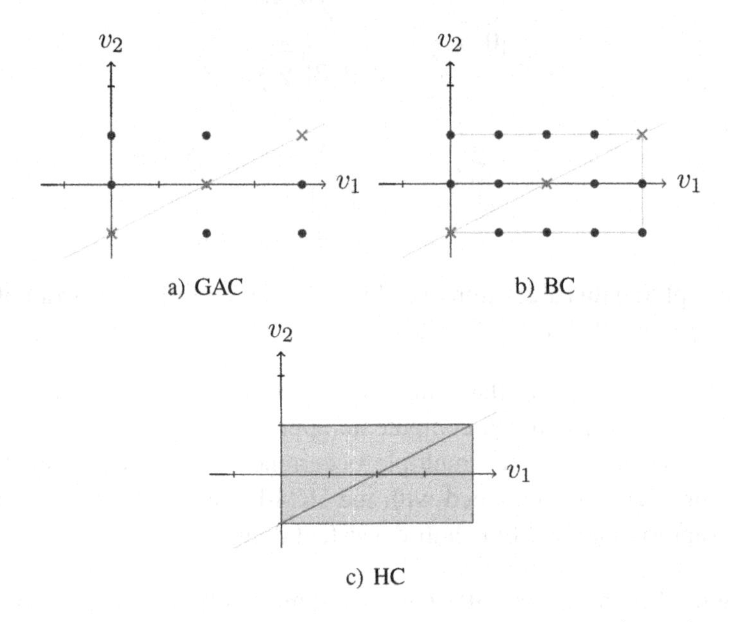


**Figure 1.5.** *Different consistencies for the constraints* $v_1 = 2(v_2 + 1)$.

For several constraints, the propagators of each constraint are iteratively applied until a fixpoint is reached. This is called the propagation loop.

1.2.2.2. *Propagation loop*

As the domains representation, previously introduced (definitions 1.10, 1.11 and 1.12), forms finite complete lattices for inclusion, it is sufficient to compute the consistency for each constraint until the fixpoint is reached. It was demonstrated by Apt in 1999 [APT 99] and Benhamou in 1996 [BEN 96] that the order in which the consistences are applied is irrelevant. Indeed, as the lattices are complete, any subset has a unique least element: the consistent fixpoint.

However, the order in which the consistencies, and therefore the propagators, are applied influences the speed of convergence, that is, the number of iterations required to reach the fixpoint. Different propagation strategies are possible. The naive strategy is to execute all the constraint propagators at each iteration. Another strategy is to execute only the propagator of the constraints containing at least one variable that has been modified in the previous iteration. This strategy is generally used. Finding an efficient algorithm computing the consistent domains for a conjunction of constraints in the general case is a real scientific challenge. In the following, the term "complexity" is used to refer to the worst-case complexity.

Since the 1970s, many algorithms have been developed for the GAC for binary constraints (AC) or for any constraints (GAC), each one improving the complexity of the previous algorithm. The first algorithm AC1 described by Mackworth in 1977 [MAC 77a] computes the arc-consistent domains for binary constraints. It verifies for each constraint ($p$), for each variable ($n$) and for each possible value (at most $d$) whether it has a support ($d^2$). This algorithm has a time complexity of $O(d^3np)$, where $d$ is the size of the largest domain, $n$ is the number of variables and $p$ is the number of constraints. This algorithm is rather basic and an improved version, called AC3, is given in the same paper [MAC 77a]. AC3 uses a list in order to only propagate the constraints in which at least one variable has been modified. AC3 has a time complexity of $O(pd^3)$ and a memory complexity of $O(p)$. The same year, an extended version of AC3, no longer restricted to binary constraint, was presented: GAC3 [MAC 77b]. This algorithm has a time complexity of $O(pk^3d^{k+1})$ and a memory complexity of $O(pk)$, with $k$ the maximum constraint arity, which is the maximum number of variables that can be in a constraint.

The algorithms AC4, which was proposed by Mohr and Henderson in 1986 [MOH 86], and GAC4, which was proposed by Mohr and Masini in 1988 [MOH 88], have a worst-case time complexity which is optimal [BES 06]. AC4 has a time and memory complexity of $O(pd^2)$. GAC4 has a time complexity of $O(pkd^k)$.

Since then, several algorithms have been designed: AC5 [HEN 92], AC6 [BES 94], AC7 [BES 99], AC2001 [BES 01] and AC3.2 [LEC 03], to name a few, ingenuously competing to reduce in practice the memory complexity and particularly the computation time. A comparison and the complexity of some of these algorithms, as well as those given above, can be found in [MAC 85] and [BES 06].

While there have been many papers and algorithms for the efficient computation of GAC, the bound consistency has been the subject of less studies. Moreover, there exist several different definitions of the bound consistency [CHO 06] (bound(D)-consistency, bound(R)-consistency and range consistency) in addition to the one considered in this book, which has changed its name several times over the years.

In the continuous variables case, the HC3 algorithm proposed in 1995 [BEN 95] computes an approximation of the HC for a conjunction of constraints. Then, in 1999 [BEN 99], the HC4 algorithm was proposed. The propagation loop applies the HC4-Revise algorithm seen earlier for each constraint containing at least one variable that has been modified. More recently, the Mohc algorithm studying the monotony of the constraints to propagate has been developed [ARA 10, ARA 12].

### 1.2.3. *Exploration*

Generally, the propagation is not sufficient to find the solutions. During the second step of the resolution process, assumptions about the variables values are made. In the case of integer variables, values are given to variables iteratively until a success (a solution is found) or a fail (a constraint is false, an empty domain) is obtained. For real variables, the idea is the same except that the domains are split into two smaller subdomains. This exploration mechanism introduces the notion of *choice point* in the resolution process.

### 1.2.4. *Resolution scheme*

The solving process alternates between phases of reduction and choice. At any point of the resolution, the propagation is first performed to remove inconsistent values from the domains as soon as possible and avoid unnecessary computations. Thus, consistency is maintained throughout the resolution. The following three cases are possible:

1) a solution is found, it is stored;

2) no solution is possible, in other words, at least one domain is empty or a constraint is false;

3) none of the above.

In the first two cases, the process returns to a previous choice point to make a new hypothesis. Several cases are possible: returning to the last choice point, called *backtracking*, or in the case of a failure, returning to the most likely point of choice responsible for the failure; in this case, it is called *backjumping* [DEC 90]. Several techniques are used to determine the node to go back to; it can be based on the dependencies between the variables or use learning during the solving process. In any case, if the search space is not yet fully explored, a choice is made, a variable and a value are chosen in the discrete case or a domain in the continuous case. This choice is taken into account by instantiating the variable or cutting the chosen domain, and then back to the first step: the propagation is performed in this new point of resolution. Thus, consistency is maintained throughout the resolution.

This solving mechanism corresponds to a search tree in which the nodes are the choice points, and the arcs correspond to a variable instantiation or a domain split.

EXAMPLE 1.18 (Search tree).– Consider a CSP on Boolean variables $v_1, v_2$ and $v_3$ on domains $D_1 = D_2 = D_3 = \{0, 1\}$ and with the constraint: $v_1$ and not $v_2$ and $v_3$. In this example, a cross corresponds to a failure and a tick corresponds to a solution:

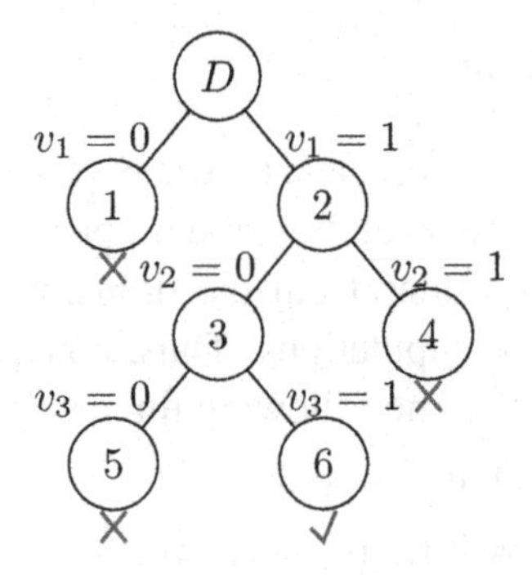


where $D = D_1 \times D_2 \times D_3$. Starting from the initial search space, the first choice is made: $v_1 = 0$. After this instantiation (node 1), the constraint is false. The exploration is stopped and goes back to the root where a new choice is made: $v_1 = 1$ (node 2). The exploration continues until all the nodes are visited. In this example, a cross corresponds to a failure and a tick corresponds to a solution.

The general scheme of a discrete solver using the GAC is shown in algorithm 1.1. Note that the branching can also be binary ($v_1 = a$ on one branch, and $v_1 \neq a$ on the other branch with $a$ a value in $D_1$). Similarly, a continuous solver is given in algorithm 1.2. It uses the HC. In these examples, the resolution stops when:

– all solutions are found, that is, when:

- all solutions have been listed in the discrete case;

- all computed boxes only contain solutions or are smaller than a certain accuracy in the continuous case.

– it has been proved that there is no solution, in other words, the search space has been fully explored without success.

Note that in both the discrete and continuous cases, the resolution process can be modified in order to stop as soon as the first solution is found.

In both solving methods, the selection criterion of the variable to instantiate or domain to cut is not explicitly given. This is because there is no unique way to choose the domain to be cut or the variable to instantiate, and it often depends on the problem to solve. The next section describes some of the choice strategies or existing exploration strategies.

### 1.2.5. *Exploration strategies*

Several choice strategies have been designed to determine the order in which the variables should be instantiated or the domains should be split. Depending on the choices made, performances may be very different. In the discrete case, the best known is perhaps the one presented in 1979 by Haralick and Elliott [HAR 79], called *first-fail*. It chooses the variable with the smallest domain. The idea is as follows: the earlier one fails, the bigger is the subtree cut in the search tree.

> *To succeed, try first where you are most likely to fail*
> –Robert M. Haralick and Gordon L. Elliott [HAR 79].

Figure 1.6 shows that the earlier the failure appears in the search tree, the bigger is the subtree cut from the search tree. In Figure 1.6(a), failure occurs later and only a small subtree in the search tree is cut. However, in Figure 1.6(b), failure occurs earlier and a larger subtree in the search tree is cut.

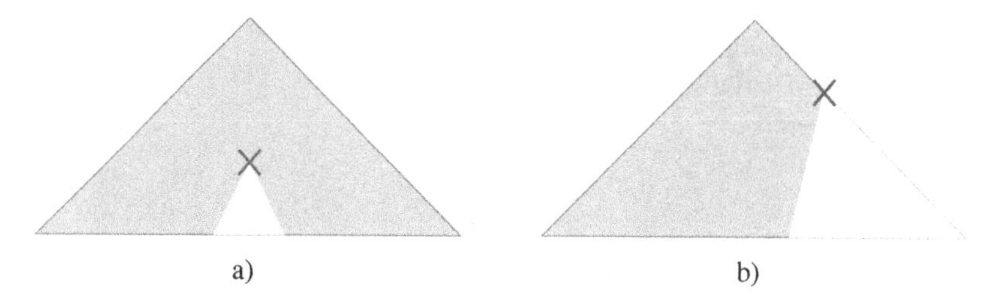


**Figure 1.6.** *Comparison of the subtree cut in the search tree given the node where the failure appears*

Figure 1.7 compares the search tree obtained by the *first-fail* strategy (Figure 1.7(b)) with that obtained by the strategy that instantiate variables with large domains first. Note that these two trees have the same area. In the case of failure, the cut in the search tree is greater when the *first-fail* strategy is used.

**Algorithm 1.1.** *A discret solver using integer Cartesian product to represent the domains. For a color version of the algorithm, see www.iste.co.uk/pelleau/abstractdomains.zip*

int $j \leftarrow 0$ /* *$j$ indicates the depth in the search tree* */
int op[] /* *at depth $j$, stores the index of the variable on which the hypothesis is made, uniformly initialized at* 0 */
int width[] /* *at depth $j$, stores the width of the tree already explored, uniformly initialized at* 0 */
integer Cartesian product $e \in \mathbb{S}$
list of integer Cartesian product sols /* *stores the solutions* */
$e \leftarrow D$ /* *initialization with the initial domains* */
**repeat**
  $e \leftarrow$ generalized arc-consistency($e$)
  width[$j$]++
  **if** $e$ is a solution **then**
    /* *success* */
    sols $\leftarrow$ sols $\cup$ $e$
  **end if**
  **if** e $= \emptyset$ or $e$ is a solution **then**
    /* *back to the last backtrackable point* */
    **while** $j \geq 0$ et width[$j$] $\geq |D_{op[j]}|$ **do**
      width[$j$] $\leftarrow 0$
      $j--$
    **end while**
  **else**
    /* *new hypothesis* */
    **choose** a variable $v_i$ to instantiate
    op[$j$] $\leftarrow i$
    $j++$
  **end if**
  **if** $j \geq 0$ **then**
    assign $v_{op[j-1]}$ to the (width[$j$] + 1)-th possible value /* *backtrackable point* */
  **end if**
**until** $j < 0$

Following this idea, the heuristic proposed in [BRÉ 79] chooses the variable with the smallest domain (dom) and appearing in the biggest number of constraints deg. In other words, the chosen variable is the one maximizing dom + deg. Then, in 1996, another strategy was presented [BES 96], choosing the variable maximizing the ratio dom/deg. In 2004 [BOU 04], a heuristic choosing the variable maximizing the ratio dom/$w$deg, with $w$ a weight associated with each constraint, was introduced. The weights are uniformly initialized to 1 at the beginning of the resolution. Then, each time a constraint is the reason of a failure, its weight is augmented. The idea is as follows: instantiate in priority variables appearing in constraints difficult to satisfy. The earlier this constraint generates a failure, the bigger is the subtree cut from the search tree. Other heuristics as a survey are given in [GEE 92, GEN 96, BEE 06].

**Algorithm 1.2.** *A classical continuous solver. For a color version of the algorithm, see www.iste.co.uk/pelleau/abstractdomains.zip*

```
list of boxes sols ← ∅                       /* stores the solutions */
queue of boxes toExplore ← ∅                 /* stores the boxes to explore */
box b ← D                                    /* initialization with the initial domains */
push b in toExplore
while toExplore ≠ ∅ do
  b ← pop(toExplore)
  b ← Hull-consistency(b)
  if b ≠ ∅ then
    if b only contains solutions or b small enough then
      /* success */
      sols ← sols ∪ b
    else
      /* new hypothesis */
      cut b in b1 and b2 by splitting along one of its dimensions
      push b1 and b2 in toExplore
    end if
  end if
end while
```

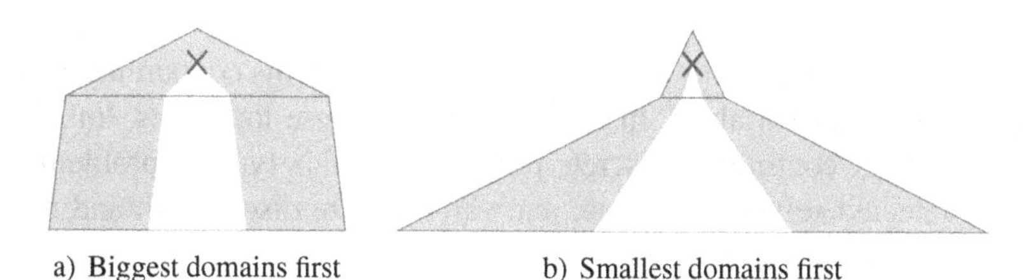

a) Biggest domains first

b) Smallest domains first

**Figure 1.7.** *Comparison between the strategy instantiating variables with the biggest domains first a) and the first-fail strategy b)*

Once the variable to instantiate is chosen, we need to choose to which value it should be instantiated. Here too, many strategies have been developed, choosing the value maximizing the number of possible solutions [DEC 87, KAS 04], the product of the domains size (*promise*) [GIN 90] and the sum of the domains size (*min-conflicts*) [FRO 95].

In the case of continuous variables, a domain is generally cut into halves. Several heuristics for splitting a domain are available: *largest-first* [RAT 94], which chooses the largest domain, thus the domains size is rapidly reduced; *round-robin,* which splits the domains one by one to ensure that all the domains are cut; and *Max-smear* [HAN 92, KEA 96], which chooses to split the domain maximizing the *smear* function of the constraints Jacobian matrix (containing the partial derivatives of each constraint). This strategy corresponds to splitting the domain of the variable with the biggest slope in the constraints.

### 1.2.6. *Discrete/continuous comparison*

It is important to note that the resolution scheme is significantly different depending on the type of variables (integer or real). In practice, the solvers are dedicated to a particular type of variable. There are many discrete solvers. Among the most known are GeCode [SCH 02], GnuProlog with constraints [DIA 12], Jacop [WOL 04], Comet [HEN 05], Eclipse [APT 07] and Minion [GEN 06]. There are much less continuous solvers in CP. The best-known solvers are Declic [BEN 97a], Numerica [HEN 97], RealPaver [GRA 06] and Ibex [CHA 09a]. Prolog IV [COL 94] and Choco 3.0 [PRU 14, FAG 14], which incorporate both a discrete and a continuous solver.

Implementation tricks are needed to solve problems containing both integer and real variables. In the following, we use the terms "mixed problem" or "continuous-discrete problem" for this type of problems. If the selected solver is discrete, real variables are discretized, and the possible values for the variables are listed for a given step.

EXAMPLE 1.19 (Discretization of a real domain).– Let $x$ be a real variable whose value lies between $0$ and $0.5$, with a step of $0.1$. Once discretized, its domain is $\{0, 0.1, 0.2, 0.3, 0.4, 0.5\}$.

This method offers the possibility to treat mixed problems but strongly depends on the chosen step. Indeed, if the chosen step is large, solutions could be lost but the variables domain is small and can easily be stored. On the contrary, if the step is too small, there are less chances of losing solutions but the domains size is very large and it explodes the combinatory of the problem. This method is used among others in the discrete solver Choco 2.0 [CHO 10]. In this case, the solver is no longer CP-complete, and it no longer guarantees that no solution is lost. But it is still CP-sound; if it returns an assignment, it is a solution.

If the selected solver is continuous, integrity constraints can be added to the problem. These new constraints specify which variables are integer and thus allow the solver to refine the bounds of these variables domains. During the propagation of these constraints, the domain bounds are rounded to the nearest integer in the proper direction. This method requires the bound consistency for the discrete variables and does not allow the GAC to be used. This method has been developed in the continuous solver RealPaver [GRA 06].

Another alternative is to add discrete global constraints to a continuous solver [BER 09] or to create mixed global constraints [CHA 09b] to treat within the same constraint both continuous and discrete variables. Thus, each variable benefits from a suitable constraint for its type (discrete or continuous). However, this method depends on the problem and demands the necessary global constraints as well as an *ad hoc* consistency for each problem.

### 1.2.7. *Conclusion*

CP can efficiently solve CSPs. While solving methods do not depend on the problem at hand, they are highly dedicated to the type of variables (discrete or continuous) of the problem. This specialization, although restrictive, makes solving methods effective. Many heuristics have been developed and used during the exploration phase to improve the results obtained in terms of both the quality of the solutions and the computation time. However, one major obstacle to the development of CP is the lack of tools and methods to solve mixed problems (with both discrete and continuous variables). For instance, there is no representation for mixing integers and reals in a single CSP while maintaining their types.

## 1.3. Synthesis

In the previous sections, we have introduced some concepts of AI and CP, two areas of research that seem to be very distant. Yet, there are similarities both in the underlying theory and in the techniques used. This section highlights some of these links and some significant differences.

### 1.3.1. *Links between AI and CP*

There are links between CP and AI; for instance, both research areas rely on the same theoretical framework (lattices and fixpoint). But, if you look closer, there are also notable differences. We present here some of the similarities and differences between these two research areas.

As previously mentioned, CP and AI rely on the same theoretical framework, and more importantly, the main goal is the same: compute an overapproximation of a desired set, expressed as solutions of constraints in CP and as the properties of a program in AI. In the particular case of discrete CSP, the approximation should be equal to the desired set. Consistency and propagation can be seen as a narrowing on abstract domains because they both help reduce the domains while remaining above the fixpoint. In AI, the narrowing operator can be used either after a widening, which ensures that the abstract domains stay larger than the desired fixpoint, or in the particular case of local iterations. In CP, domains are always reduced and the propagation is used to reduce the domains as soon as possible.

Note that in both cases, it is not allowed to reach the fixpoint. For example, in CP, for certain constraints, such as the global constraint `nvalue`, computing the consistency is NP-hard [BES 04] and it is allowed to compute an overapproximation of the consistent domains even for discrete constraints.

REMARK 1.16.– On the theoretical framework, another significant difference between CP and AI should be noted. In AI, fixed-point theorems are fundamental and the basis of loops analysis, a key step in the analysis of a program. In AI, the lattices are usually infinite, while in CP, lattices used for consistencies are always finite. They are built from sets starting bounded (initial domains) and their components must be computer representable (integers or floating point), so in a finite number.

These two research areas rely on the computation of fixpoint or overapproximations of fixpoints. However, the techniques used to compute them are significantly different. First, in CP, while solving a given problem, for a series of choice points, the approximations are strictly decreasing, except when the fixpoint is reached. While in AI, the approximations can increase, for instance, when approximating a loop. Second, CP aims at completeness by improving solutions using refinement, while AI generally embraces incompleteness.

Furthermore, the static analyzer precision is conditioned by the chosen abstract domain and the used operators. When a refinement is performed, domains become more precise, which can be done by either restarting the analyzer with the obtained domains [CLA 03], or manually, which requires changing the analyzer [BER 10]. Another way to improve the accuracy is to add local iterations [GRA 92]. To sum up, the accuracy of an abstract domain is not defined as such because it is implied by the choice made for the analysis (abstract domains, transfer function, widening, narrowing, etc.). In contrast, CP integrates for continuous domains an explicit definition of accuracy. The choice of abstract domain in a CP solver does not change its precision (which is fixed), but its efficiency (the amount of computation needed to achieve the desired accuracy).

Another significant difference is that all the techniques in AI are parameterized by abstract domains. On the contrary, CP solvers are highly dependent on the variables type (integer or real) and are especially dedicated to one domain representation. There exists no CP solver for which the domains representation is a parameter; in particular, there are no mixed solvers. In section 1.2, we saw that in CP there are only three representations for the variables domain given the type of the variable. On the contrary, in AI, an important research effort has been made on abstract domains. This has led to the development of many Cartesian or relational domains, such as polyhedra, octagons and ellipsoid, to name a few. Note that they can depend on the variable type, like the binary decision diagrams for binary variables [BER 10]; but more importantly, they can analyze the different variables type within the same analyzer.

### 1.3.2. *Analysis*

CP allows us to model a wide variety of problems under a same and unique form, the CSP. It offers a wide variety of constraints, including, in the case of discrete variables, more than 300 global constraints. They can be used to describe complex relationships between the variables and come with efficient propagation algorithms. A number of studies have focused on improving the complexity, and therefore the efficiency of these algorithms. Another strong point of CP is that it provides generic solving methods that do not depend on the problem to solve, and a single constraint can be used in different problems. In order to improve these solving methods performances, many heuristics have been developed, such as the variable or value choice heuristic for the instantiation, or the choice heuristic for the domain to split. However, CP offers very few representations for the domains, and the solving methods are restricted to only one type of variables (discrete or continuous).

On the other side, AI provides a large number of representations, abstract domains. These can be of different shapes (box, polyhedron, ellipse, etc.) and are not defined for a particular type of variables. Thus, they may represent both discrete and continuous variables. Moreover, using several abstract domains allows the analyzers to effectively deal with very large programs, containing a large number of variables and lines of code. Another advantage of abstract domains is that they come with efficient algorithms for the widening and narrowing operators to compute in a small number of iterations an approximation of the fixpoint. However, this approximation is not always very precise, even after applying the narrowing operator or local iterations.

To sum up, CP offers very few representations for the domains but efficient algorithms, while in AI there are very rich representations for the domains and algorithms to compute overapproximations but no solving algorithm (in the sense that the accuracy is not easily configurable).

There have already been works at the frontier between AI and CP. For instance, CP has been used to verify programs [COL 07], to analyze characteristic models and automatically generate configuration tests [HER 11b], to verify constraint satisfaction problems model [LAZ 12] or even to improve the results obtained by a static analyzer [PON 11].

Another approach for program verification is to use a satisfiability solver [KIN 69]. Recent years have witnessed significant improvements in the methods of Boolean satisfiability problem (SAT) and satisfiability modulo theories (SMT) [KRO 08], as well as in their applications (e.g., the C bounded model checker (CBMC) [KRO 14]). Furthermore, D'Silva *et al.*, [D'SI 12] have recently proposed to express the SAT algorithms in the context of AI (using fixpoints and abstractions), a promising way for cross-pollination between these two research areas.

The work presented in this book is similar, except that we are at the intersection of AI and CP. There are of course similarities between the solving process in CP and in SAT/SMT; however, both the model and the solving methods differ. Resolution methods in SAT and SMT used a model based on Boolean variables, for which the algorithms are dedicated. CP combines the constraints on any type of variables and sometimes loses efficiency to gain expressiveness.

# 2

# Abstract Interpretation for the Constraints

In this chapter we give unified definitions for the different components of the resolution process in constraint programming (CP). With these new definitions, a unique resolution method can be defined for any abstract domain. This new resolution scheme no longer depends on the type of the variables. Moreover, the definition of the abstract domains in CP gives the possibility of solving problems using non-Cartesian domain representations.

## 2.1. Introduction

In CP, solving techniques strongly depend on the type of variables, and are even dedicated to a type of variables (integer or real). If a problem contains both integer and real variables, there are no solving methods. There are three possible solutions to this kind of problem with CP: the integers are transformed into real variables and integrity constraints are added to the problem (the propagation of these new constraints refine the bound of the integer variables [GRA 06]); the real varibales are discretized, the possible values for the real variables are enumerated with a given step [CHO 10]; or a discrete and a continuous solver are combined [COL 94, FAG 14]

By looking more closely, we can see that regardless of the resolution method used, it alternates between propagation and exploration phases. We can even go further and say that these two methods differ on only three points: the representation of the domains, the consistency used and the way domains are cut. The splitting operator used during the solving process strongly depends on the chosen representation. Indeed, we need to know some characteristics of the chosen representation, such as the size to cut into smaller elements. Similarly, the consistency used strongly depends on the chosen representation. In fact, the generalized arc consistency would not be used if the domains are represented using integer intervals. If domains are represented using integer intervals, the bound consistency is used. If the integer Cartesian product is used, then the generalized arc consistency is more appropriate. And if the domains are represented using floating point intervals, the hull consistency is more suitable. Therefore, the consistency does not depend only on the domain representation, but also on the constraints.

Inspired by abstract domains in abstract interpretation (AI), we define abstract domains CP as an object that contains, among other things, a computer representation, a splitting operator and a function to compute its size. The consistency could be added to the abstract domains, but we prefer to dissociate it since it does not depend only on the domain representation. With this definition of abstract domains in CP, we obtain a unique solving method that no longer depends on the type of variables, or the domain representation. In this new solving method, the abstract domain is a parameter. In addition, we are no longer restricted to existing Cartesian representations, but can define new representations in the same way as in AI.

## 2.2. Unified components

To begin with, we define all the necessary bricks for the development of a unique solving method, namely, the consistency, the splitting operator and of course the abstract domains for CP. These definitions are based on notions of order, lattices and fixpoints.

### 2.2.1. *Consistency and fixpoint*

Given a partially ordered set with the inclusion and a constraint, we can define the consistent element as the least element of the partially ordered set if it exists. Similarly, for a conjunction of constraints, the set of least elements for each constraint forms a partially ordered set with the inclusion, and the consistent element is therefore defined as the least element of this set. As a consequence, from this observation, we have the following definitions and propositions.

Let $E$ be a subset of $\mathcal{P}(\hat{D})$, where $\hat{D}$ is the initial search space, with the inclusion as a partial order. This set corresponds to the chosen representation for the domains. We will write $E^f$ for $E\backslash\emptyset$. In the following, we will restrict the definitions to cases where $E$ is closed by intersection, as this is sufficient for classical cases. Note that, if $E$ is closed by intersection then it possesses the greatest lower bound and is a directed lattice.

DEFINITION 2.1 (E-consistency).– *Consider a constraint $C$. An element e is $E$-consistent for $C$ if and only if it is the least element of $E$ containing all the solutions for $C$, $S_C$. In other words, it is the least element of:*

$$\mathbb{C}_C^E = \{e' \in E, S_C \subseteq e'\}$$

All the main consistencies defined in section 1.2.2, whether discrete or continuous, are included in this definition. The following propositions combine each existing consistency with its corresponding subset. These propositions ensure that the main consistencies are properly included in definition 2.1.

PROPOSITION 2.1.– Let $\mathbb{S}$ be the set of Cartesian products of finite subsets of integers. The $\mathbb{S}$-consistency is the generalized-arc-consistency (GAC) (definition 1.15).

PROOF 2.1.–

*GAC* $\Longrightarrow$ $\mathbb{S}$*-consistent*

We first prove that GAC $\Longrightarrow$ $\mathbb{S}$-consistent. Let $D = D_1 \times ... \times D_n$ GAC for a constraint $C$. $D$ obviously contains all the solutions, and we need to prove that it is the smallest such element of $\mathbb{S}$. Let $D' \in \mathbb{S}$ strictly smaller than $D$. Then there is an $i$ such that $D'_i \subset D_i$ and a $v \in D_i \setminus D'_i$. Since $v \in D_i$ and $D$ is GAC, there also exist $x_j \in D_j$ for $j \neq i$ such that $(x_1 \dots v \dots x_n)$ is a solution. This solution is not in $D'$ and thus any $D' \subset D$ loses solutions. $\diamond$

*$\mathbb{S}$-consistent $\Longrightarrow$ GAC*

We now prove that $\mathbb{S}$-consistent $\Longrightarrow$ GAC. Let $D = D_1 \times \cdots \times D_n$ $\mathbb{S}$-consistent for a constraint $C$. Let $i \in [\![1, n]\!]$ and $v \in D_i$. Suppose that for all other $x_j \in D_j$, $(x_1 \dots v \dots x_n)$ is not a solution. We can construct the set $D_1 \times \cdots \times (D_i \setminus \{v\}) \times \cdots \times D_n$ strictly smaller than $D$ containing all the solutions. Hence $D$ is not the smallest element. There can be no such $i$ and $v$, and $D$ is GAC. $\diamond$

A domain is GAC if and only if it is $\mathbb{S}$-consistent. $\square$

PROPOSITION 2.2.– Let $\mathbb{IB}$ be the set of integer boxes (Cartesian product of finite interval of integers). The $\mathbb{IB}$-consistency is the bound-consistency (BC) (definition 1.16).

PROOF 2.2.–

*BC $\Longrightarrow$ $\mathbb{IB}$-consistent*

We first prove that BC $\Longrightarrow$ $\mathbb{IB}$-consistent. Let $D = D_1 \times \cdots \times D_n$ BC for a constraint $C$. $D$ obviously contains all the solutions, and we need to prove that it is the smallest such element of $\mathbb{IB}$. Let $D' \in \mathbb{IB}$ be strictly smaller than $D$. Then there is an $i$ such that $D'_i \subset D_i$. Let $v$ be one of the bound of $D_i$ such that $v \notin D'_i$. Since $D$ is BC, there also exists $x_j \in D_j$ for $j \neq i$ such that $(x_1 \dots v \dots x_n)$ is a solution. This solution is not in $D'$ and thus any $D' \subset D$ loses solutions. $\diamond$

*$\mathbb{IB}$-consistent $\Longrightarrow$ BC*

We now prove that $\mathbb{IB}$-consistent $\Longrightarrow$ BC. Let $D = D_1 \times \cdots \times D_n$ $\mathbb{IB}$-consistent for a constraint $C$. Let $D_i = [\![a_i, b_i]\!]$. Suppose that for all other $x_j \in D_j$, $(x_1 \dots a_i \dots x_n)$ is not a solution. We can construct the set $D_1 \times \cdots \times [\![a_i + 1, b_i]\!] \times \dots D_n$ strictly smaller that $D$ containing all the solutions. Hence $D$ is not the smallest element. There can be no such $i$ and $v$, and $D$ is BC. ◇

A domain is BC if and only if it is $\mathbb{IB}$-consistent. □

PROPOSITION 2.3.– Let $\mathbb{B}$ be the set of boxes. The $\mathbb{B}$-consistency is the hull consistency (HC) (definition 1.17).

PROOF 2.3.– *HC* $\Longrightarrow$ $\mathbb{B}$*-consistent*

We first prove that HC $\Longrightarrow$ $\mathbb{B}$-consistent. Let $D = D_1 \times \cdots \times D_n$ HC for a constraint $C$. $D$ obviously contains all the solutions, and we need to prove that it is the smallest such element of $\mathbb{B}$. Let $D' \in \mathbb{B}$ strictly smaller than $D$. Then there is an $i$ such that $D'_i \subset D_i$. Let $I \in D_i \setminus D'_i$. Since $I \in D_i$ and $D$ is HC, there also exists $I_j \in D_j$ for $j \neq i$ such that $C(I_1, \ldots, I, \ldots, I_n)$. This solution is not in $D'$ and thus any $D' \subset D$ loses solutions. ◇

$\mathbb{B}$*-consistent* $\Longrightarrow$ *HC*

We now prove that $\mathbb{B}$-consistent $\Longrightarrow$ HC. Let $I \in \mathbb{B}$, $\mathbb{B}$-consistent for a constraint $C$. Then $I$ is the least element of $\mathbb{C}_C^{\mathbb{B}}$ which corresponds to the smallest Cartesian product of intervals with floating-point bounds containing all the solutions for $C$. Hence, using definition 1.17, $I$ is HC. ◇

A domain is HC if and only if it is $\mathbb{B}$-consistent. □

Definition 2.1 of $E$-consistency thus generalizes the existing consistency. For any $E \subseteq \mathcal{P}(\hat{D})$ closed by intersection, consistency is well-defined. From this definition of $E$-consistency the following proposition is derived:

PROPOSITION 2.4.– If $E$ is closed under infinite intersection, $\mathbb{C}_C^E$ is a complete lattice and there exists a unique $E$-consistent element for $C$ in $E$. If it exists, this element is written as $\mathbf{C}_C^E$.

REMARK 2.1.– Note that if $E$ is not closed under intersection, then the least element of $\mathbb{C}_C^E$ does not always exist. For instance, in the particular case where $C$ is a circle and $E$ is the set of convex polyhedra, there exists no smallest polyhedron containing all the solutions. This is due to the fact that polyhedra are not closed by intersection.

PROOF 2.4.– Let $\mathbf{C}_C^E = \bigcap_{e \in E, S_C \subseteq e} e$.

*Least elements*

We first prove that $\mathbf{C}_C^E$ is the least element of $\mathbb{C}_C^E$. Let $A \in \mathbb{C}_C^E$ strictly smaller than $\mathbf{C}_C^E$, we have $S_C \subseteq A$. Since $\mathbf{C}_C^E$ is the intersection of all the elements containing the solutions $S_C$, $A$ cannot be smaller than $\mathbf{C}_C^E$. Thus, $\mathbf{C}_C^E$ is the least element of $\mathbb{C}_C^E$. ◇

*Unicity*

We now prove that $\mathbf{C}_C^E$ is unique. Suppose that there is an element $B \in \mathbb{C}_C^E$ such that $B$ is also a least element of $\mathbb{C}_C^E$. If such an element exists, then it also contains all the solutions, $S_C \subseteq B$. Since $\mathbf{C}_C^E$ is the intersection of all the elements containing the solutions $S_C$, we have $\mathbf{C}_C^E \subseteq B$. Hence $\mathbf{C}_C^E$ is unique. ◇

$\mathbf{C}_C^E$ is unique and the least element of $\mathbb{C}_C^E$. □

Definition 2.1 can easily be extended for a conjunction of constraints as follows:

DEFINITION 2.2 ($E$-consistency for a conjunction of constraints).– *Let $e \in E$, $e$ is $E$-consistent for $C_1 \ldots C_p$ (or a subset) if and only if it is a least element for $\mathbb{C}_{C_1 \wedge \cdots \wedge C_p}^E = \{e' \in E, S_{C_1 \wedge \cdots \wedge C_p} \subseteq e'\}$. If such an element exists, it is written as $\mathbf{C}_{C_1 \wedge \cdots \wedge C_p}^E$. If there is no ambiguity $\mathbb{C}_{C_1 \wedge \cdots \wedge C_p}^E$ is written as $\mathbb{C}^E$ and its least element as $\mathbf{C}^E$.*

This definition will preferably be used in the case where $E$ is closed under intersection.

PROPOSITION 2.5.– If $E$ is closed by intersection, then $\mathbf{C}^E$ exists and is unique. In addition, the set of all $\mathbf{C}_{C_{i_1} \wedge \cdots \wedge C_{i_k}}^E$ for $i_1 \ldots i_k \in [\![1, p]\!]$ forms a lattice for inclusion and $\mathbf{C}_{C_1 \wedge \cdots \wedge C_p}^E$ is its least element.

PROOF 2.5.– *Unicity*

Unicity of $\mathbf{C}^E$ directly comes from proposition 2.4. ◇

*Complete lattice*

We first prove that the set of all $\mathbf{C}^E_{C_{i_1}\wedge\dots\wedge C_{i_k}}$ for $i_1 \dots i_k \in [\![1,p]\!]$ forms a lattice for inclusion. Let $\mathbf{C}^E_{C_i}$ and $\mathbf{C}^E_{C_j}$ for $i, j \in [\![1,p]\!]$ be any two elements of the set. Then the pair $\{\mathbf{C}^E_{C_i}, \mathbf{C}^E_{C_j}\}$ has both a greatest lower bound

$$\mathbf{C}^E_{C_i} \cap \mathbf{C}^E_{C_j} = \bigcap_{e\in E, S_{C_i}\subset e \vee S_{C_j}\subset e} e$$

and a least upper bound

$$\mathbf{C}^E_{C_i} \cup \mathbf{C}^E_{C_j} = \bigcap_{e\in E, S_{C_i}\subset e, S_{C_j}\subset e} e$$

Thus, any pair $\{\mathbf{C}^E_{C_i}, \mathbf{C}^E_{C_j}\}$ has a *glb* and a *lub*. Hence, the set of all $\mathbf{C}^E_{C_{i_1}\wedge\dots\wedge C_{i_k}}$ for $i_1 \dots i_k \in [\![1,p]\!]$ is a lattice. ◇

*Least element*

We now prove that $\mathbf{C}^E_{C_i\wedge C_j}$ is the least element for $\mathbf{C}^E_{C_i}, \mathbf{C}^E_{C_j}$, that is $\mathbf{C}^E_{C_i\wedge C_j}$ is included in $\mathbf{C}^E_{C_i} \cap \mathbf{C}^E_{C_j}$. Let us first prove that $\mathbf{C}^E_{C_i\wedge C_j} \subseteq \mathbf{C}^E_{C_i}$. As $S_{C_i\wedge C_j} = \{(s_1 \dots s_n) \in D, C_i(s_1 \dots s_n) \wedge C_j(s_1 \dots s_n)\}$ and $S_{C_i} = \{(s_1 \dots s_n) \in D, C_i(s_1 \dots s_n)\}$, then $S_{C_i\wedge C_j} \subseteq S_{C_i}$. Thus $\bigcap_{e\subseteq E, S_{C_i\wedge C_j}\subseteq e} e \subseteq \bigcap_{e\subseteq E, S_{C_i}\subseteq e} e$. Hence $\mathbf{C}^E_{C_i\wedge C_j} \subseteq \mathbf{C}^E_{C_i}$. Similarly, we prove that $\mathbf{C}^E_{C_i\wedge C_j} \subseteq \mathbf{C}^E_{C_j}$. As $\mathbf{C}^E_{C_i\wedge C_j} \subseteq \mathbf{C}^E_{C_i}$ and $\mathbf{C}^E_{C_i\wedge C_j} \subseteq \mathbf{C}^E_{C_j}$, then $\mathbf{C}^E_{C_i\wedge C_j} \subseteq \mathbf{C}^E_{C_i} \cap \mathbf{C}^E_{C_j}$. Therefore $\mathbf{C}^E_{C_i\wedge C_j}$ is the least element for $\mathbf{C}^E_{C_i}, \mathbf{C}^E_{C_j}$. ◇

□

This proposition provided that each constraint of the CSP comes with a propagator, it is sufficient to apply these propagators iteratively, no matter the order, until the fixpoint is reached, to achieve consistency for the CSP. On existing consistencies, a similar idea has been proposed in [APT 99] or in [BEN 96].

EXAMPLE 2.1 ($\mathbb{B}$-consistency).– Consider the CSP on real variables $v_1, v_2$ on domains $D_1 = [0, 5]$, $D_2 = [0, 5]$ and the constraints

$C1$: $5v_1 + v_2 \geq 10$
$C2$: $2v_1 + 5v_2 \leq 20$
$C3$: $2v_2 - v_1 \geq 1$

Figure 2.1 shows the lattice with inclusion of this CSP $\mathbb{B}$-consistencies. For each element in this lattice, we have in dark pink the set of solutions and in light pink the set approximated by the $\mathbb{B}$-consistency represented by a green box. The dotted black boxes are only there to frame the elements. For the elements corresponding to intersections, the dashed boxes correspond to the boxes in the intersection and the result is a plain box.

We can see that $\forall i_1, \ldots, i_k \in [\![1, 3]\!], \mathbf{C}^{\mathbb{B}}_{C_{i_1} \wedge \cdots \wedge C_{i_k}} \subseteq \mathbf{C}^{\mathbb{B}}_{C_{i_1}} \cap \cdots \cap \mathbf{C}^{\mathbb{B}}_{C_{i_k}} \subseteq \mathbf{C}^{\mathbb{B}}_{C_{i_1}}, \ldots, \mathbf{C}^{\mathbb{B}}_{C_{i_k}}$. For instance, we have $\mathbf{C}^{\mathbb{B}}_{C_1 \wedge C_3} \subseteq \mathbf{C}^{\mathbb{B}}_{C_1} \cap \mathbf{C}^{\mathbb{B}}_{C_3} \subseteq \mathbf{C}^{\mathbb{B}}_{C_1}, \mathbf{C}^{\mathbb{B}}_{C_3}$.

### 2.2.2. *Splitting operator*

Once the consistent element is computed, we still need to explore it in order to find the solutions. This is done by cutting the remaining search space into smaller elements. An element is cut using a splitting operator as defined below.

DEFINITION 2.3 (Splitting operator in $E$).– *Let $(E, \subseteq)$ be a partially ordered set. A* splitting operator *is an operator $\oplus : E \to \mathcal{P}(E)$ such that $\forall e \in E$,*

1) $|\oplus(e)|$ *is finite*, $\oplus(e) = \{e_1, \ldots, e_k\}$,

2) $\bigcup_{i \in [\![1,k]\!]} e_i = e$,

3) $\forall i \in [\![1, k]\!], e \neq \emptyset \Longrightarrow e_i \neq \emptyset$,

4) $\exists i \in [\![1, k]\!], e_i = e \Longrightarrow$ *e is a least element of $E^f$.*

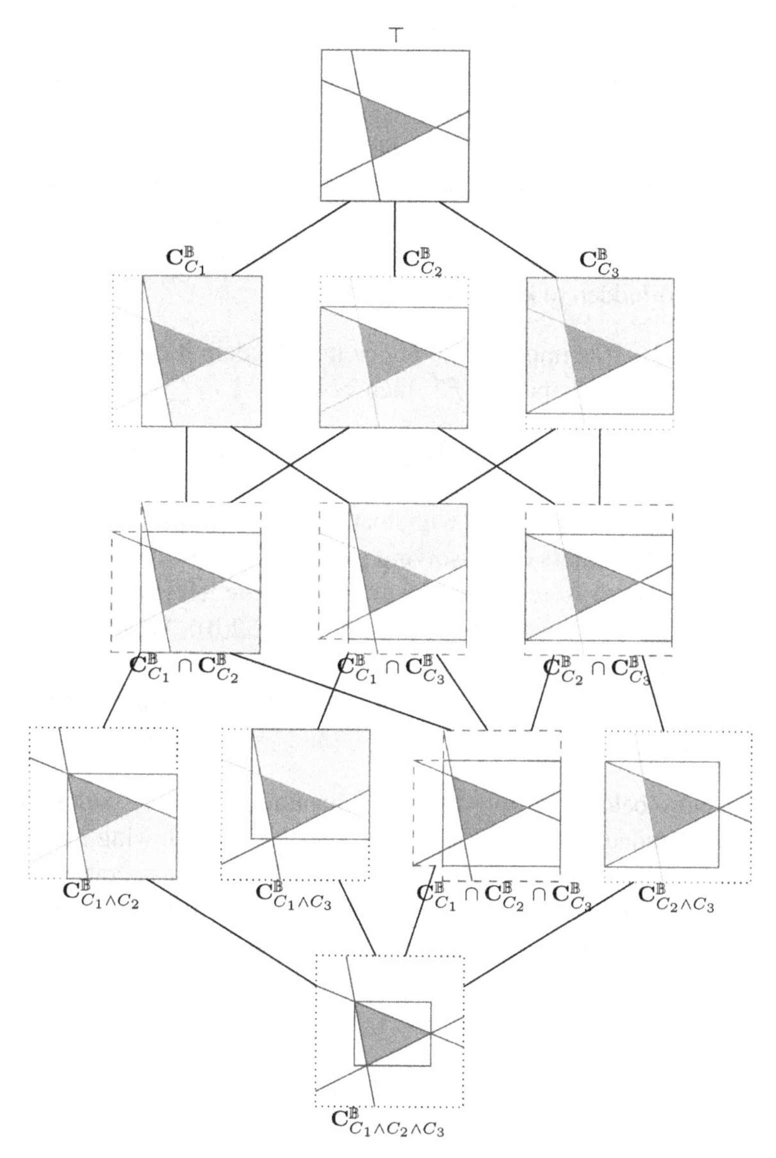


**Figure 2.1.** *Example of finite lattices for the inclusion, the elements of which are the $\mathbb{B}$-consistencies of the CSP given in example 2.1. For a color version of the figure, see www.iste.co.uk/pelleau/abstractdomains.zip*

The first condition is needed for the search process to terminate. Indeed, the width of the search tree must be finite so that the search tree can be. The second condition ensures that the solving process is complete; in other words, the splitting operator does not loose solutions. Moreover, this condition ensures that no element is added. The third condition forbids the splitting operator to return an empty domain. The fourth condition means that the splitting operator actually splits: it is forbidden to keep the same domains.

REMARK 2.2.– It is important to notice that the definition implies that if $e$ is not the least element of $E^f$, then $\forall i \in [\![1, k]\!], e_i \subsetneq e$.

REMARK 2.3.– Also note that this definition of the splitting operator does not ask for $\oplus(e)$ to be a partition of $e$. This is mandatory in order to include the split on intervals with floating-point bounds. This does not affect the completeness of the solving process. As shown in Figure 2.2, a splitting operator is applied to the element on the left (Figure 2.2(a)), giving us the two elements on the right (Figure 2.2(b)). We can see that the starting element is included in the union of the elements obtained with the splitting operator and that the intersection of these elements is not empty.

We show below that the discrete instantiation and the continuous split are included in the definition. We will use the following notation for Cartesian domains: let $\oplus_1 : E_1 \rightarrow \mathcal{P}(E_1)$ be an operator for a partially ordered set $E_1$. Let $E_2$ be another partially ordered set and $Id$ the identity function on $E_2$. We write $\oplus_1 \times Id$, the operator on $E_1 \times E_2$ such that $\oplus_1 \times Id(e_1, e_2) = \bigcup_{e \in \oplus_1(e_1)} e \times e_2$. We also write $Id^i$ for the Cartesian product of $i$ times $Id$.

EXAMPLE 2.2 (Integer instantiation).– Instantiation of a discrete variable is a splitting operator on $\mathcal{P}(\mathbb{N})$: $\oplus_{\mathbb{N}}(d) = \bigcup_{v \in d}\{v\}$. For every $i \in [\![1, n]\!]$, the operator $\oplus_{\mathbb{N}^n, i}(d) = Id^{i-1} \times \oplus_{\mathbb{N}} \times Id^{n-i-1}$, where $\oplus_{\mathbb{N}}$, at the $i$-th place, is a splitting operator. This consists of choosing a variable $v_i$ and a value $v$ in $D_i$.

EXAMPLE 2.3 (Splitting operator for an interval).– The continuous splitting operator is $\mathcal{P}(\mathbb{I}) : \oplus_{\mathbb{I}}(I) = \{I^{\vdash}, I^{\dashv}\}$ where $I^{\vdash}$ and $I^{\dashv}$ are two non-empty subintervals of $I$ such that $I^{\vdash} \cup I^{\dashv} = I$ with whatever way of managing the rounding errors on floats. The most usual splitting operator for continuous CSP, for $I = [a, b]$, returns $I^{\vdash} = [a, h]$ and $I^{\dashv} = [h, b]$ with $h \in \mathbb{F}, h = \frac{a+b}{2}$ rounded in any direction. To ensure that the splitting operator terminates, it stops when $a$ and $b$ are two consecutive floats.

EXAMPLE 2.4 (Splitting operator for a Cartesian product of intervals).– The usual splitting operator for a Cartesian product of intervals is defined in $\mathcal{P}(\mathbb{B})$. Let $I \in \mathbb{B}, D = I_1 \times \cdots \times I_n$. The splitting operator first chooses an interval $I_i, i \in [\![1, n]\!]$. Then this interval is split using $\oplus_{\mathbb{I}}$. Thus, we have:

$$\oplus_{\mathbb{B}}(I) = \{I_1 \times \cdots \times I_i^{\vdash} \times \cdots \times I_n, I_1 \times \cdots \times I_i^{\dashv} \times \cdots \times I_n\}$$

where $\{I_i^{\vdash}, I_i^{\dashv}\} = \oplus_{\mathbb{I}}(I_i)$. The most common continuous splitting operator in CP, generally splits the domain maximizing $\max_{i \in [\![1,n]\!]}(\overline{I_i} - \underline{I_i})$.

In the following, we will write, for any $E$, $\oplus_E$ is the splitting operator in $E$. Due to these generic definitions of consistency and splitting operator, we can now define abstract domains in CP.

### 2.2.3. *Abstract domains*

Given that our goal is to define a generic solver totally independent of the domain representation, we define in this section, the abstract domains for CP. These are defined such that they have the requirements to be part of a solving process.

DEFINITION 2.4 (Abstract domain for CP).– *An* abstract domain *is defined by:*

– *a complete lattice $E$;*

– *a concretization $\gamma : E \to D$, and an abstraction $\alpha : D \to E$ forming a Galois connection between E and the search space;*

– *a computer representable normal form;*

– *a sequence of splitting operators for $E$;*

– *a monotonic size function* $\tau : E \rightarrow \mathbb{R}^+$ *such that* $\tau(e) = 0 \Leftrightarrow e = \emptyset$.

The Galois connection connects the abstract domain to a computer representation of the CSP domains. However, as for the abstract domains in AI, it may not exist. The normal form ensures that there exists a unique representation for each abstract domain and that they can be comparable. The size function $\tau$ represents some measure of the abstract domain. It is an artificial technique to express the precision of an abstract domain. It is used later as a termination criterion in the solving method.

Abstract domains can be defined independently from the domains of a particular CSP. They are intended to represent the shape of the domain representation. Of course, they can be Cartesian, but this is no longer mandatory. Note that we do not formalize the propagators as a part of an abstract domain. As they depend both on the constraints and on the shape of the abstract domain, they have to be *ad hoc*.

With this definition, we can, for instance, define the Shadok[1] abstract domain by assuming that we can find a way to compute it, that it is computer representable and that there exists a consistency calculating, if it exists, the smallest Shadok containing the solutions. Figure 2.2(b) shows the result of a possible splitting operator.

As the abstract domains are now defined for CP, a solving method based on abstract domains can be defined.

## 2.3. Unified solving

With the definitions of consistency (definition 2.1) and a splitting operator (definition 2.3) that no longer depend on the domain representation, and with the definition of abstract domains in CP (definition 2.4), it is now possible to define a unified solving method based on abstract domains. This method is defined as the classical

1 http://www.lesshadoks.com/ or http://en.wikipedia.org/wiki/Les_Shadoks.

iteration of propagations and splits, but is no longer dedicated to a representation.

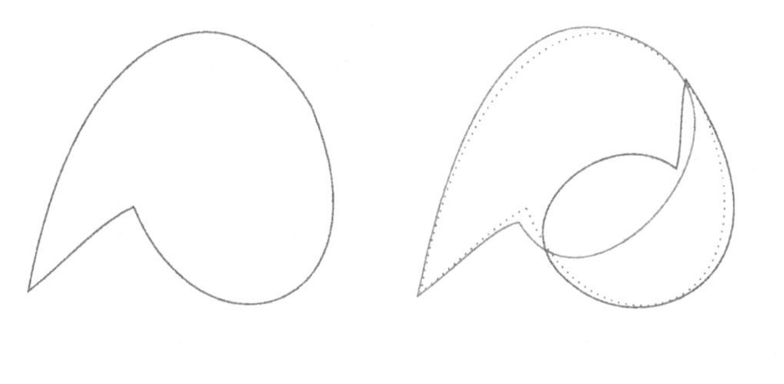

a) a Shadok b) an example of a Shadok split

**Figure 2.2.** *We can define a Shadok abstract domain. A Shadok in its normal form is given a) as well as the result of a possible splitting operator b). For a color version of the figure, see www.iste.co.uk/pelleau/abstractdomains.zip*

Algorithm 2.1 gives the pseudo-code of a solving method based on abstract domains. This algorithm looks a lot like the one for the continuous solving method (algorithm 1.2). Indeed, if we replace everywhere in the pseudo-code "abstract domain" by "box", the continuous solver is obtained. The same applies for the integer Cartesian product.

The following proposition gives the hypothesis under which this algorithm is complete and terminates.

PROPOSITION 2.6.– If $E$ is closed by intersection (H1), has no infinite decreasing chain (H2), and if $r \in \tau(E^f)$ (H3), then the solving process in Figure 2.1 terminates and is complete, in the sense that no solution is lost.

PROOF 2.6.–

*Termination*

We first prove that this algorithm terminates. Suppose that the search tree is infinite. Using definition 2.3 of the splitting operator, the width of the search tree is finite. Thus, it exists as an infinite branch. Along this branch, abstract domains are strictly decreasing as long as $e$ is not the least element of $E^f$. Using hypothesis (H2), there is a $K$ such that $\forall k \geq K, e_k = e_K$. Let us study the different possible cases:

– if $e_K = \emptyset$, then the algorithm terminates as $e_K$ contains no solution;

– if $e_K \neq \emptyset$ and $(\tau(e_K) \leq r$ or $e_K \subset S)$, then the algorithm terminates as $e_K$ is a solution or is small enough compared to $r$;

– if $e_K \neq \emptyset$ and $\tau(e_K) > r$, then $e_K$ is split and $e_K \neq e_{K+1}$ which contradicts the definition of $K$.

We can thus conclude that the solving process in algorithm 2.1 terminates. ◇

*Completeness*

As the splitting operator (definition 2.3) and the consistency (definition 2.1) are complete, the algorithm 2.1 is complete. ◇

Under hypothesis (H1), (H2) and (H3), the solving algorithm 2.1 terminates and is complete. □

**Algorithm 2.1.** *Solving with abstract domains. For a color version of the algorithm, see www.iste.co.uk/pelleau/abstractdomains.zip*

```
list of abstract domains sols ← ∅            /* stores the abstract solutions */
queue of abstract domains toExplore ← ∅ /* stores the abstract elements to
explore */
abstract domain e ∈ E          /* data structure for the abstract domain */

e = α(D̂)                          /* initialization with initial domains */
push e in toExplore

while toExplore ≠ ∅ do
  e ← pop(toExplore)
  e ← E-consistance(e)
  if e ≠ ∅ then
    if τ(e) ≤ r or e ⊆ S then
      sols ← sols ∪ e
    else
      choose a splitting operator ⊕_E
      push ⊕_E(e) in toExplore
    end if
  end if
end while
```

This proposition defines a generic solver on any abstract domain $E$ with some hypotheses. Indeed, hypotheses (H1) and (H2) must be true, and $r$ well chosen. If $r$ is too large, the abstract element returned as solutions by the algorithm will be too large and can contain a large number of non-solution points. On the contrary, if $r$ is too small, the solving can take a long time and even worse, in the case of integer Cartesian products, given the precision function $\tau$, solutions may not be found. The efficiency of this solver will of course depend on the consistency algorithms on $E$ and on the computer representation of the chosen abstract domain.

The usual solving algorithms in CP are included in this definition as shown below.

EXAMPLE 2.5 (Solving with integer Cartesian products).– For a fixed $n$, the set $\mathbb{S}$ with the splitting operator $\oplus_{\mathbb{N}^n,i}$ and the precision $\tau_{\mathbb{S}}(e) = \max(|X_i|)$ for $i \in [\![1, n]\!]$ is an abstract domain satisfying (H1) and (H2). In order to model the fact that the search process ends when all the variables are instantiated, we can take $r = 1$. As mentioned earlier, the $\mathbb{S}$-consistency is the generalized arc-consistency.

EXAMPLE 2.6 (Solving with integer boxes).– For a fixed $n$, the set $\mathbb{IB}$ with the splitting operator $\oplus_{\mathbb{N}^n,i}$ and the precision $\tau_{\mathbb{IB}}(e) = \max(b_i - a_i)$ for $i \in [\![1, n]\!]$ is an abstract domain satisfying (H1) and (H2). In order to model the fact that the search process ends when all the variables are instantiated, we can take $r = 1$. As mentioned earlier, the $\mathbb{IB}$-consistency is the bound-consistency.

EXAMPLE 2.7 (Solving with intervals).– For a fixed $n$, the set $\mathbb{B}$ with the splitting operator $\oplus_{\mathbb{B},i}$ and the precision $\tau_{\mathbb{B}}(I) = \max(\overline{I_i} - \underline{I_i})$ for $i \in [\![1, n]\!]$ is an abstract domain satisfying (H1) and (H2). In order to model the fact that the search process ends when a precision $r$ is reached, we can stop when $\tau_{\mathbb{B}} \leq r$. Solving with $\mathbb{B}$ corresponds to the generally used solving process in continuous solvers with hull consistency.

These three examples show how to retrieve the usual abstract domains, which are Cartesian and of a single type (integer Cartesian products or intervals). It is now possible to define an abstract solver as an alternation of propagations and splits, see Figure 2.1. This solver returns an array of abstract domains containing only solution ($e \subseteq S$) or whose precision is less than $r$. The union of all of these elements is an approximation of the solution set.

## 2.4. Conclusion

In this chapter, we showed that it is possible to define a solver that is independent of the domains representation and therefore of the variables type of the problem. This solver includes usual CP solving methods, regardless of the type of variables (discrete or continuous). Furthermore, the domains are no longer limited to Cartesian representations. With this new definition, it is now possible to use different representations for domains. There exists a large number of abstract domains in AI, such as the polyhedra, ellipsoids and zonotopes to name a few. In the next chapter, we describe a non-Cartesian representation: the octagons.

# 3

# Octagons

In the generic solver presented in the previous chapter, abstract domains can be used to solve constraint satisfaction problems. In the same way as in abstract interpretation (AI), we define the octagon abstract domain for constraint programming (CP). This abstract domain already exists in AI, and was introduced by Miné [MIN 06]. We first define a representation that is computer representable, a splitting operator and a precision function. Then, a Galois connection between the octagons and the boxes is given.

## 3.1. Definitions

In geometry, an octagon is, in $\mathbb{R}^2$, a polygon with eight sides[1]. As part of this work, we use a more general definition that was introduced in [MIN 06]. In the following, the term "octagon" is used to denote the octagons as defined below, when we want to talk about octagons as described in geometry, the term "mathematical octagon" will be used.

DEFINITION 3.1 (Octagonal constraint).– *Let* $v_i, v_j$ *be two variables. We call* octagonal constraints *constraints of the form* $\pm v_i \pm v_j \leq c$ *with* $c \in \mathbb{R}$ *a constant.*

1 http://mathworld.wolfram.com/Octagon.html.

REMARK 3.1.– Interval constraints ($v_i \geq a$, $v_i \leq b$) are particular cases of octagonal constraints.

REMARK 3.2.– In a conjunction of constraints, two octagonal constraints are said to be redundant if and only if they have the same left side. For instance, the following constraints $v_1 - v_2 \leq c$ and $v_1 - v_2 \leq c'$ are redundant. Only one of these constraints is effective, i.e. the one with the smallest constant (the smallest right side).

For instance, in $\mathbb{R}^2$, octagonal constraints define straight lines which are parallel to the axis if $i = j$ and diagonal if $i \neq j$. This stays true in $\mathbb{R}^n$, where octagonal constraints define half-spaces (cut by hyperplanes). Consider a cube in $\mathbb{R}^3$, by adding octagonal constraints ($\pm v_i \pm v_j \leq c$), the edges are going to be cut, but not the corners.

DEFINITION 3.2 (Octagon).– *An octagon is the set of points in $\mathbb{R}^n$ satisfying a conjunction of octagonal constraints.*

REMARK 3.3.– The geometric shape defined above includes the octagons but also other polygons. For instance, in $\mathbb{R}^2$, an octagon can have less than eight sides. In general, in $\mathbb{R}^n$, an octagon has at most $2n^2$ faces, which is the maximum number of possible non-redundant octagonal constraints on $n$ variables. Moreover, octagons satisfying a conjunction of octagonal constraints are necessarily convex.

Also, note that an octagon is a set of *real* points, but, like for the intervals, they can be restricted to having floating-point bounds ($c \in \mathbb{F}$). We thus have a real octagon with floating-point bounds.

An example of an octagon in $\mathbb{R}^n$ is given in Figure 3.1. We can see that an octagon can only have seven sides.

The first part of Figure 3.2 is composed of examples of octagons (Figure 3.2(a)). We can see that an octagon does not necessarily have eight sides in two dimensions, and that all the sides are parallel to the axis or to the diagonals. The second part (Figure 3.2(b)) shows the polygons that do not correspond to the definition of octagons given previously (definition 3.2). The first polygon has all its sides parallel to the axis or to the diagonals but is not convex. The next two examples are mathematical octagons but their sides are not parallel to the axis or the diagonals, and they do not correspond to definition 3.2.

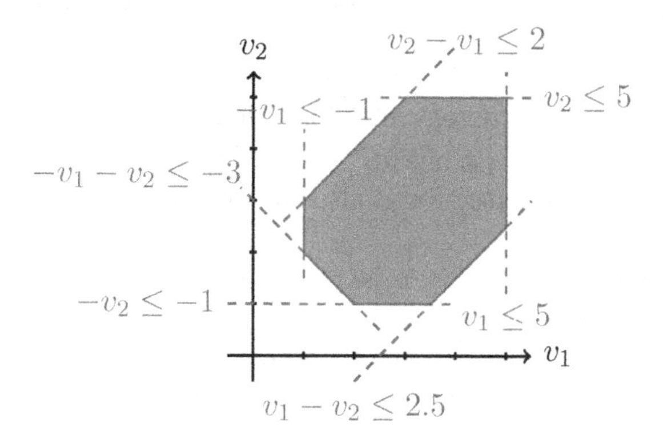


**Figure 3.1.** *Example of an octagon in* $\mathbb{R}^2$

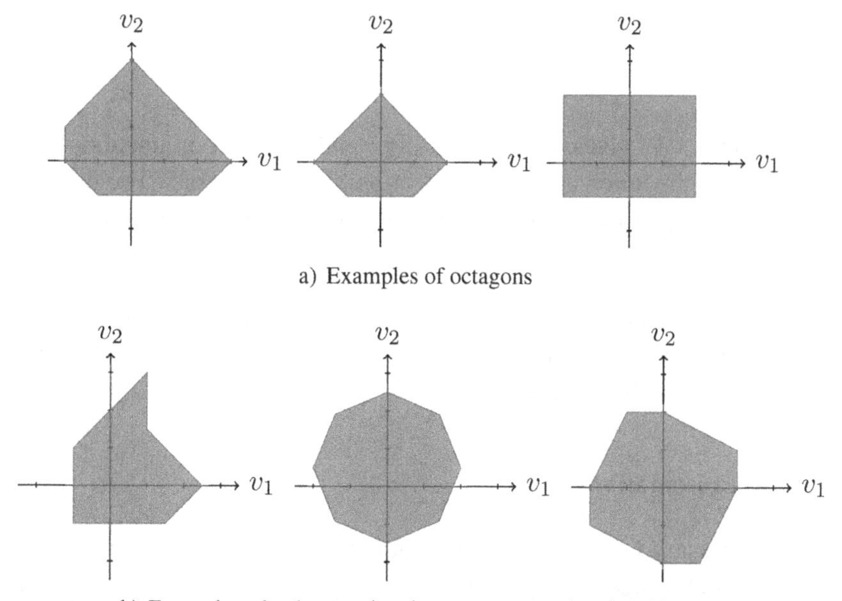


**Figure 3.2.** *Examples of octagons a) and polygons that do not respect the definition of octagons b)*

This definition of octagons offers interesting properties, which do not exist for mathematical octagons, such as the closure under intersection. Indeed, the intersection of any two octagons is still an octagon.

REMARK 3.4.– The octagons are closed under intersection. Consider any two octagons $\mathbf{O} = \{\pm v_i \pm v_j \leq c\}$ and $\mathbf{O}' = \{\pm v_i \pm v_j \leq c'\}$ for $i, j \in [\![1, n]\!]$. Their intersection is also an octagon:

$$\mathbf{O} \cap \mathbf{O}' = \{\pm v_i \pm v_j \leq \min(c, c')\}$$

Octagons are not closed under union, but the smallest octagon containing the union of any two octagons can easily be computed.

REMARK 3.5.– Consider any two octagons $\mathbf{O} = \{\pm v_i \pm v_j \leq c\}$ and $\mathbf{O}' = \{\pm v_i \pm v_j \leq c'\}$ for $i, j \in [\![1, n]\!]$. The smallest octagon including the union is:

$$\mathbf{O} \cup \mathbf{O}' = \{\pm v_i \pm v_j \leq \max(c, c')\}$$

Figure 3.3 gives, geometrically and with conjunctions of constraints, the union and the intersection of two octagons. The dashed octagon corresponds to the smallest octagon including the union. The dotted octagon corresponds to the intersection.

In the following, octagons are restricted to octagons with floating-point bounds ($c \in \mathbb{F}$). Without loss of generality, we assume octagons to be defined with no redundancies.

## 3.2. Representations

A necessary feature of abstract domains is that their elements must be computer representable. There exists no unique way to represent an abstract domain; however, some representations are better suited for certain types of computations. Here, we provide two possible representations for octagons. The first one, the matrix representation, was introduced in 1983 by Menasche and Berthomieu in [MEN 83]. It is used in [MIN 06]. The second one, the box representation, is part of the contribution of this work and is more suited for some computations in CP, such as the consistency.

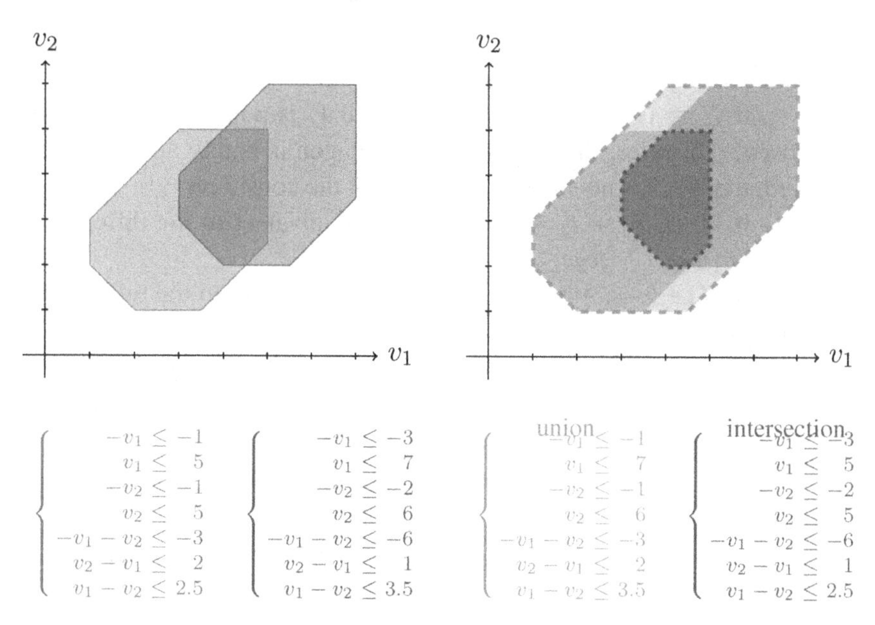


**Figure 3.3.** *Example of the union and the intersection of two octagons. The union gives the dashed orange octagon ant the intersection gives the dotted octagon. For a color version of the figure, see www.iste.co.uk/pelleau/abstractdomains.zip*

### 3.2.1. *Matrix representation*

An octagon can be represented with a *difference bound matrix* (DBM) as described in [MEN 83, MIN 06]. This representation is based on a normalization of the octagonal constraints as follows.

DEFINITION 3.3 (Difference constraint).– *Let $w, w'$ be two variables. The* difference constraint *is a constraint of the form $w - w' \leq c$ with $c \in \mathbb{F}$ a constant.*

By introducing new variables, it is possible to rewrite an octagonal constraint as an equivalent difference constraint. Let $C \equiv (\pm v_i \pm v_j \leq c)$ be an octagonal constraint. New variables $w_{2i-1}, w_{2i}$ are introduced such that $w_{2i-1}$ corresponds to the positive form of $v_i$ and $w_{2i}$ to the negative form of $v_i$. In other words, $\forall i \in [\![1, n]\!], w_{2i-1} = v_i$ and $w_{2i} = -v_i$. Then:

– for $i = j$

- if $C \equiv (v_i - v_i \leq c)$, then, $c \geq 0, C$ is pointless and can be removed; otherwise, the corresponding octagon is empty and we stop. Indeed, there exists no value of $v_i$ verifying the constraint $v_i - v_i < 0$;

- if $C \equiv (v_i + v_i \leq c)$, then $C$ is equivalent to the difference constraint $(w_{2i-1} - w_{2i} \leq c)$;

- if $C \equiv (-v_i - v_i \leq c)$, then $C$ is equivalent to the difference constraint $(w_{2i} - w_{2i-1} \leq c)$;

– for $i \neq j$

- if $C \equiv (v_i - v_j \leq c)$, then $C$ is equivalent to the difference constraints $(w_{2i-1} - w_{2j-1} \leq c)$ and $(w_{2j} - w_{2i} \leq c)$;

- if $C \equiv (v_i + v_j \leq c)$, then $C$ is equivalent to the difference constraints $(w_{2i-1} - w_{2j} \leq c)$ and $(w_{2j-1} - w_{2i} \leq c)$;

- if $C \equiv (-v_i - v_j \leq c)$, then $C$ is equivalent to the difference constraints $(w_{2i} - w_{2j-1} \leq c)$ and $(w_{2j} - w_{2i-1} \leq c)$;

- if $C \equiv (-v_i + v_j \leq c)$, then $C$ is equivalent to the difference constraints $(w_{2i} - w_{2j} \leq c)$ and $(w_{2j-1} - w_{2i-1} \leq c)$.

In the following, original variables are denoted by $(v_1 \ldots v_n)$ and the corresponding new variables by $(w_1, w_2, \ldots w_{2n})$ with $w_{2i-1} = v_i$ and $w_{2i} = -v_i$. Miné [MIN 06] showed that the difference constraints created by replacing positive and negative occurrences of the variable $v_i$ by the corresponding $w_i$ represent the same octagon as we obtained with octagonal constraints. Difference constraints can be stored in a difference bound matrix and octagonal constraints can be translated into difference constraints; hence, a difference bound matrix is a possible representation for octagons.

DEFINITION 3.4 (Difference bound matrix).– *Let* $\mathbf{O}$ *be an octagon in* $\mathbb{F}^n$*, and its set of difference constraints as defined previously. The* difference bound matrix *is a square matrix* $2n \times 2n$ *such that the element line* $i$ *and column* $j$ *is the constant* $c$ *from the difference constraint* $w_j - w_i \leq c$.

An example is given in Figure 3.4(b) in which the octagon is given graphically (Figure 3.4(a)) and corresponds to the difference bound matrix (Figure 3.4(b)). The element on line 1 and column 3 and the element on line 4 and column 2 correspond to the same constraint: $v_2 - v_1 \leq 2$.

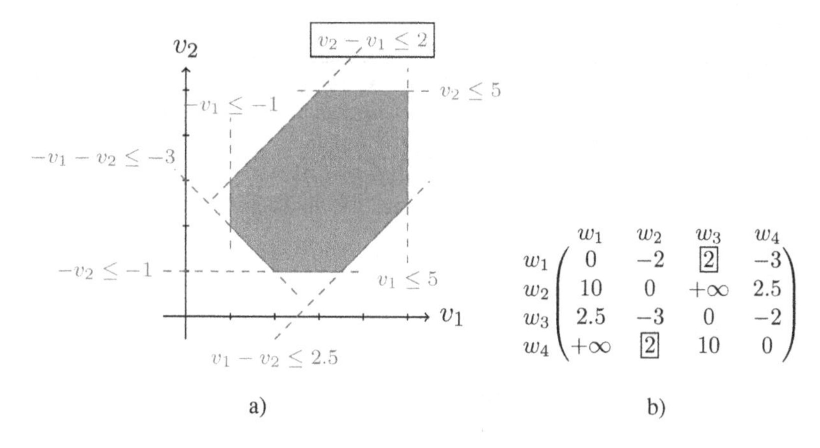


**Figure 3.4.** *Equivalent representations of an octagon: with octagonal constraints a) and with a difference bound matrix b)*

At this stage, different difference bound matrices may represent the same octagon. For example, in Figure 3.4(b), the element on line 2 and column 3 can be replaced by 100 without changing the corresponding octagon. In [MIN 06], an algorithm is defined to optimally compute the smallest values for the difference bound matrix. This algorithm is an adaptation of the Floyd–Warshall shortest path algorithm [FLO 62]. It is modified in order to take advantage of the difference bound matrix structure. It exploits the fact that $w_{2i-1}$ and $w_{2i}$ correspond to the same variable.

Algorithm 3.1 gives the pseudocode of the modified version of the Floyd–Warshall shortest path algorithm for octagons. The instruction $i' \leftarrow (i \bmod 2 = 0)\,?\,i-1 : i+1$ signifies: if $i$ is even, then $i'$ takes the value $i - 1$; otherwise, $i'$ takes the value $i + 1$. The beginning of this algorithm corresponds to the Floyd–Warshall (computation of the shortest path part). The rest has been added and exploits the fact that

$w_{2i-1}$ and $w_{2i}$ correspond to the same variable $v_i$. Moreover, elements on the main diagonal must be greater than or equal to zero; otherwise, an error is raised because the corresponding octagon is empty. Indeed, $\forall i, w_i - w_i \leq c$, $c$ cannot be less than 0. Another version of algorithm 3.1 is proposed in [BAG 09]. These two versions have the same complexity ($O(n^3)$).

**Algorithm 3.1.** *Modified version of Floyd–Warshall shortest path algorithm for the octagons. For a color version of the algorithm, see www.iste.co.uk/pelleau/abstractdomains.zip*

```
float dbm[2n][2n]          /* square difference bound matrix of 2n × 2n */
int i, j, k, i', j'              /* indexes for the difference bound matrix */

for k from 1 to n do
   for i from 1 to 2n do
      for j from 1 to 2n do
         /* Computation of the shortest path */
         dbm[i][j] ← min( dbm[i][j], dbm[i][2k] + dbm[2k][j],
                          dbm[i][2k − 1] + dbm[2k − 1][j],
                          dbm[i][2k − 1] + dbm[2k − 1][2k] + dbm[2k][j],
                          dbm[i][2k] + dbm[2k][2k − 1] + dbm[2k − 1][j] )
      end for
   end for

   /* Added to the original version */
   for i from 1 to 2n do
      for j from 1 to 2n do
         i' ← (i mod 2 = 0) ? i − 1 : i + 1
         j' ← (j mod 2 = 0) ? j − 1 : j + 1
         dbm[i][j] ← min(dbm[i][j], dbm[i][i'] + dbm[j'][j])
      end for
   end for
end for
for i from 1 to 2n do
   if dmb[i][i] < 0 then
      return error
   else
      dbm[i][i] ← 0
   end if
end for
```

The execution of this modified version of the Floyd–Warshall on the difference bound matrix given in Figure 3.4(b) replaces both $+\infty$ by 10, which corresponds to adding the constraint $v_1 + v_2 \leq 10$ at the conjunction of octagonal constraints. Note that the implementation of the Floyd–Warshall algorithm replaces the $+\infty$ by 12.

We introduce a new representation for octagons based on the boxes (definition 1.12). This representation combined with a difference bound matrix will be used to define, from an initial set of continuous constraints, an equivalent system taking into account the octagonal domains.

### 3.2.2. *Intersection of boxes representation*

In two dimensions, an octagon can be represented by the intersection of one box in the canonical basis for $\mathbb{F}^2$, and one box in the basis obtained from the canonical basis by a rotation of angle $\pi/4$. Figure 3.5 gives the representation by intersection of boxes (Figure 3.5(b)) corresponding to the octagon as illustrated in Figure 3.5(a). In this example, the octagon is defined by seven non-redundant octagonal constraints, one of the boxes is thus unbound.

To generalize this notion to bigger dimensions, we introduce the notion of rotated basis as follows.

DEFINITION 3.5 (Rotated basis).– *Let* $B = (u_1, \dots, u_n)$ *be the canonical basis of* $\mathbb{F}^n$. *Let* $\alpha = \pi/4$. *The* (i,j)-rotated basis $B_\alpha^{i,j}$, *for* $i, j \in [\![1, n]\!]$, *is the basis obtained after a rotation of* $\alpha$ *in the subplane defined by* $(u_i, u_j)$, *and the other vectors remaining unchanged:*

$$B_\alpha^{i,j} = (u_1, \dots, u_{i-1}, (\cos(\alpha)u_i + \sin(\alpha)u_j), \dots$$
$$u_{j-1}, (-\sin(\alpha)u_i + \cos(\alpha)u_j), \dots u_n)$$

By convention, for any $i \in [\![1, n]\!]$, $B_\alpha^{i,i}$ represents the canonical basis. In what follows, $\alpha$ is always $\pi/4$ and will be omitted. Finally, for $i, j, k \in [\![1, n]\!]$, every variable $v_k$ living in the $B^{i,j}$-rotated basis and whose domain is $D_k$ will be denoted by $v_k^{i,j}$ and its domain by $D_k^{i,j}$.

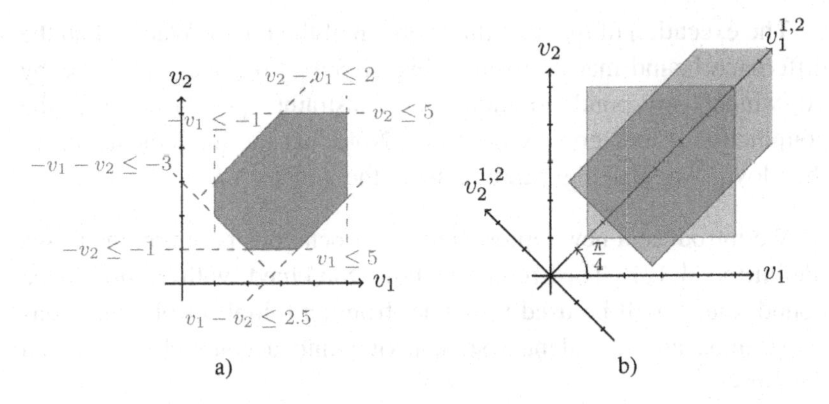


**Figure 3.5.** *Equivalent representations for the same octagon: the octagonal constraints a) and the intersection of boxes b). For a color version of the figure, see www.iste.co.uk/pelleau/abstractdomains.zip*

REMARK 3.6.– Note that the idea of rotating variables and constraints has already been proposed in [GOL 10] in order to better approximate the solution set. However, their method is dedicated to underconstrained systems of equations.

The difference bound matrix can also be interpreted as a representation of the intersection of one box in the canonical basis and $n(n-1)/2$ other boxes, each living in a rotated basis.

Let $\mathbf{O}$ be an octagon in $\mathbb{F}^n$ and $M$ its difference bound matrix, with the same notations as above ($M$ is a $2n \times 2n$ square matrix). For $i, j \in [\![1, n]\!]$, with $i \neq j$, let $\mathbf{B}_{\mathbf{O}}^{i,j}$ be the box $I_1 \times \cdots \times I_i^{i,j} \times \cdots \times I_j^{i,j} \cdots \times I_n$, in the basis $B^{i,j}$, such that $\forall k \in [\![1, n]\!]$

$$\begin{aligned}
I_k &= [\ \underline{-\tfrac{1}{2}M[2k-1,2k]}\ ,\ \overline{\tfrac{1}{2}M[2k,2k-1]}\ ] \\
I_i^{i,j} &= [\ \underline{-\tfrac{1}{\sqrt{2}}M[2j-1,2i]}\ ,\ \overline{\tfrac{1}{\sqrt{2}}M[2j,2i-1]}\ ] \\
I_j^{i,j} &= [\ \underline{-\tfrac{1}{\sqrt{2}}M[2j-1,2i-1]}\ ,\ \overline{\tfrac{1}{\sqrt{2}}M[2j,2i]}\ ]
\end{aligned}$$

EXAMPLE 3.1 (From the difference bound matrix to boxes).– Considering the difference bound matrix in Figure 3.4(b), the boxes are

$I_1 \times I_2 = [1,5] \times [1,5]$ and $I_1^{1,2} \times I_2^{1,2} = \left[\underline{3/\sqrt{2}}, +\infty\right] \times \left[\underline{-2.5/\sqrt{2}}, \overline{\sqrt{2}}\right]$, which correspond to the boxes given in Figure 3.5(b).

PROPOSITION 3.1.– Let $\mathbf{O}$ be an octagon in $\mathbb{F}^n$, and $\mathbf{B}_{\mathbf{O}}^{i,j}$ the boxes as defined above. Then, $\mathbf{O} = \bigcap_{i,j \in [\![1,n]\!]} \mathbf{B}_{\mathbf{O}}^{i,j}$.

PROOF 3.1.– Let $i, j \in [\![1, n]\!]$. We have $v_i^{i,j} = \frac{v_i + v_j}{\sqrt{2}}$ and $v_j^{i,j} = \frac{v_j - v_i}{\sqrt{2}}$ by definition 3.5. Thus, $(v_1 \ldots v_i^{i,j} \ldots v_j^{i,j} \ldots v_n) \in \mathbf{B}_{\mathbf{O}}^{i,j}$ if and only if it satisfies the octagonal constraints on $v_i$ and $v_j$, and the unary constraints for the other coordinates, in the difference bound matrix. The box $\mathbf{B}_{\mathbf{O}}^{i,j}$ is thus the solution set for these particular octagonal constraints. The points in $\bigcap_{i,j \in [\![1,n]\!]} \mathbf{B}_{\mathbf{O}}^{i,j}$ are exactly the points which satisfy all the octagonal constraints. □

To summarize, there is a different way to represent an octagon. An octagon can be represented by a difference bound matrix interpreted as a set of octagonal constraints (definition in intension) or equivalently as an intersection of boxes (definition in extension). Moreover, the conversion from one representation to the other is at the cost of a multiplication/division with the appropriate rounding mode.

We now have computer representable forms for the octagons. To define the octagon abstract domain, we need to define some operators, such as the splitting operator and the size function for the octagons. In the following, we assume that the octagons are closed by the modified Floyd–Warshall algorithm, and that they are only defined by non-redundant octagonal constraints.

## 3.3. Abstract domain components

In the second phase of the solving process (exploration), if the current abstract element is not considered as a solution, a splitting operator is used to cut the current abstract element into smaller abstract elements. To solve using octagons, a splitting operator is necessary, as well as a precision function to compute the size of an octagon. To define these operators, we use the intersection of box representation. Naturally, these operators can be equivalently defined on the difference bound matrix.

### 3.3.1. *Octagonal splitting operator*

The octagonal splitting operator defined here extends the usual split operator to octagons. Splits can be performed in the canonical basis, thus being equivalent to the usual splits, or in a rotated basis. It can be defined as follows.

DEFINITION 3.6 (Octagonal splitting operator).– *Let* $\mathbf{O}$ *be an octagon defined as the intersection of boxes* $I_1, \ldots, I_n, I_1^{1,2}, \ldots, I_1^{n-1,n}, \ldots, I_n^{n-1,n}$*, such that* $\forall i, j, k \in [\![1, n]\!], I_k^{i,j} = [a, b]$ *with* $a, b \in \mathbb{F}$*. The splitting operator* $\oplus_o(\mathbf{O})$ *for variable* $v_k^{i,j}$ *computes the two octagonal subdomains* $I_1, \ldots, [a, h], \ldots, I_n^{n-1,n}$ *and* $I_1, \ldots, [h, b], \ldots, I_n^{n-1,n}$ *with* $h = \frac{a+b}{2}$ *rounded in* $\mathbb{F}$.

It is easily verified that the union of the two octagonal subdomains is equal to the original octagon. Thus, this splitting operator does not lose solutions and corresponds to definition 2.3. Note that this operator has three parameters $i, j$ and $k$ and not only one parameter $i$ like in the interval splitting operator. However, this definition does not take into account the correlation between the variables of the different basis. We again take advantage of the octagonal representation to communicate the domain reduction to the other basis. A split is thus immediately followed by a Floyd–Warshall propagation to update the octagonal constraints.

Figure 3.6 shows an example of the octagonal splitting operator. The octagon on the left is split along the red dashed line.

### 3.3.2. *Octagonal precision*

In most continuous solvers, the precision is defined as the size of the largest domain. For octagons, this definition leads to a loss of information because it takes the domains separately and does not take into account the correlation between the variables. In other words, this definition does not take into account that variables $v_1^{\{1,2\}}$ and $v_2^{\{1,2\}}$ depend on $v_1$ and $v_2$. Thus, we define an octagonal precision function taking the correlations of variables into account and corresponding to the diameter on all axes and diagonals.

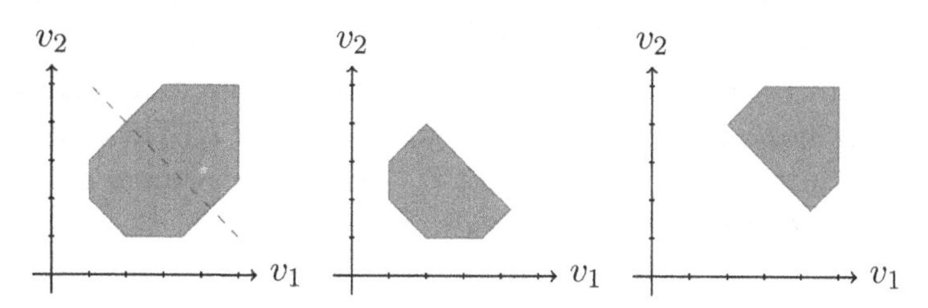


**Figure 3.6.** *Example of a split: the octagon on the left is cut in the $B^{1,2}$ basis. For a color version of the figure, see www.iste.co.uk/pelleau/abstractdomains.zip*

DEFINITION 3.7 (Octagonal precision).– *Let* **O** *be an octagon with its box representation $I_1 \dots I_n, I_1^{1,2} \dots I_n^{n-1,n}$. The octagonal precision is:*

$$\tau_o(\mathbf{O}) = \min_{i,j \in [\![1,n]\!]} \left( \max_{k \in [\![1,n]\!]} \left( \overline{I_k^{i,j}} - \underline{I_k^{i,j}} \right) \right)$$

For a single regular box, $\tau_o$ would be the same precision as usual. On an octagon, we take the minimum precision of the boxes in all the bases because it is more accurate. Moreover, this definition allows us to retrieve the operational semantics of the precision, as shown by the following proposition: in an octagon of precision $r$ overapproximating a solution set $S$, every point is at a distance at most $r$ from $S$.

PROPOSITION 3.2.– Consider a constraint satisfaction problem on variables $(v_1 \dots v_n)$, with domains $(\hat{D}_1 \dots \hat{D}_n)$, and constraints $(C_1 \dots C_p)$. Let **O** be an octagon overapproximating the solution set $S$ of this problem and containing at least one solution $s$. Let $r = \tau_o(\mathbf{O})$. Let $(x_1, \dots, x_n) \in \mathbb{F}^n$ be a point in $I_1 \times \dots \times I_n$, $\forall i \in [\![1,n]\!], I_i \subseteq \hat{D}_i$. Then, $\forall i \in [\![1,n]\!]$, $\min_{s \in S} |v_i - s_i| \leq r$, where $s = (s_1 \dots s_n)$. Each coordinate of all the points in **O** is at a distance at most $r$ from a solution.

PROOF 3.2.– Using definition 3.7, the precision $r$ is the minimum of some quantities in all the rotated bases. Let $B^{i,j}$ be the basis that realizes this minimum. The box $\mathbf{B}_{\mathbf{O}}^{i,j} = I_1 \times \cdots \times I_i^{i,j} \times \cdots \times I_j^{i,j} \cdots \times I_n$ is hull consistent and thus contains all the solutions $S$. Let $s \in S$, because $r = max_k(\overline{I_k} - \underline{I_k}), \forall k \in [\![1, n]\!], |s_k - v_k| \leq \overline{I_k} - \underline{I_k} \leq r$. Hence, any point is at a distance at most $r$ from a solution. □

Figure 3.7 shows an example of the octagonal precision. Consider the octagon obtained with the intersections of the canonical box (in blue) and the box in the rotated basis (in pink). The usual precision function on boxes returns for this octagon the size of $I_1$. The octagonal precision function returns the size of $I_1^{1,2}$ which seems more adequate for the given octagon.

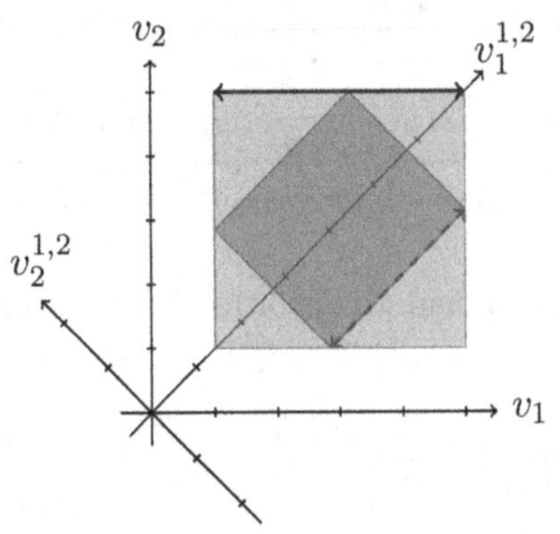


**Figure 3.7.** *Example of the octagonal precision: the usual continuous precision returns the size of $I_1$ (the blue plain double arrow) while the octagonal precision returns the size of $I_1^{1,2}$ (the red dashed double arrow). For a color version of the figure, see www.iste.co.uk/pelleau/abstractdomains.zip*

To define the octagon abstract domain in CP, it only remains to prove that all octagons form a complete lattice and define a Galois connection between the octagons and floating-point bound boxes.

## 3.4. Abstract domains

In this section, we first show that the set of octagons forms a complete lattice with inclusion.

PROPOSITION 3.3.– The set of octagons forms a complete lattice $\mathbb{O}$.

PROOF 3.3.– Let $\mathbf{O}$ and $\mathbf{O}'$ be any octagons. The set $\{\mathbf{O}, \mathbf{O}'\}$ has both a *lub* $\mathbf{O} \cup \mathbf{O}'$ (remark 3.4) and a *glb* $\mathbf{O} \cap \mathbf{O}'$ (remark 3.5). Thus, any finite set has both a *lub* and a *glb*. Hence, $\mathbb{O}$ forms a complete lattice with inclusion. □

We now show that there exists a Galois connection between octagons and floating-point bound boxes.

PROPOSITION 3.4.– There exists a Galois connection $\mathbb{O} \underset{\alpha_o}{\overset{\gamma_o}{\leftrightarrows}} \mathbb{B}$.

PROOF 3.4.–

– $\alpha_o : \mathbb{O} \rightarrow \mathbb{B}, \mathbf{O} \mapsto \{\pm v_i \leq c, i \in [\![1, n]\!]\}$ with $\mathbf{O} = \{\pm v_i \pm v_j \leq c\}$: only the constraints of the form $\pm v_i \pm v_i \leq c$ are kept, which correspond to the box bounds;

– $\gamma_o : \mathbb{B} \rightarrow \mathbb{O}, \mathbf{B} \mapsto \mathbf{B}$: is straightforward and exact as a box is an octagon.

The abstraction function $\alpha_o$ and the concretization function $\gamma_o$ form a Galois connection. □

From propositions 3.3 and 3.4, we can say that octagons can be the base set of an abstract domain.

DEFINITION 3.8 (Octagon abstract domain).– *The octagon abstract domain is defined by:*

– *the complete lattice* $\mathbb{O}$*;*

– *the Galois connection* $\mathbb{O} \underset{\alpha_o}{\overset{\gamma_o}{\leftrightarrows}} \mathbb{B}$*;*

– *the representation with intersection of boxes;*

– *the octagonal splitting operator* $\oplus_o$*;*

– *the octagonal precision function* $\tau_o$*.*

REMARK 3.7.– In this definition, we have chosen the representation with intersection of boxes, but it can be replaced with the matrix representation as they are equivalent.

This definition of octagons relies on the generation of all the possible bases $\frac{n(n+1)}{2}$. This transforms a constraint satisfaction problem (CSP) with $n$ variables into a representation with $n^2$ variables. Among all these bases, some are perhaps less relevant than others. We present in the following a partial definition of the octagons:

DEFINITION 3.9 (Partial octagon).– *Let $\mathcal{J}, \mathcal{K} \in \mathcal{P}([\![1, n]\!])$ be sets of indexes. Let* **O** *be a set of octagonal constraints* $\{\pm v_i \pm v_j \leq c \,|\, i \in \mathcal{J}, j \in \mathcal{K}\} \cup \{\pm v_i \pm v_i \leq c \,|\, i \in [\![1, n]\!]\}$. *The subset of $\mathbb{R}^n$ points satisfying all the constraints in* **O** *is called a partial octagon.*

EXAMPLE 3.2 (Partial octagon).– Let $(v_1, v_2, v_3)$ be a set of variables. Let $\mathcal{J} = \{1\}$ and $\mathcal{K} = \{2, 3\}$. The partial octagon corresponds to the set of points satisfying the following set of octagonal constraints $\{\pm v_1 \pm v_2 \leq c, \pm v_1 \pm v_3 \leq c'\} \cup \{\pm v_i \pm v_i \leq c \,|\, i \in [\![1, n]\!]\}$.

The definition of partial octagons is very similar to the octagons. In both cases, it is a set of points satisfying a set of octagonal constraints. However, in the partial case, only a subset of all the possible non-redundant octagonal constraints is expressed. Note that this definition includes the octagons. Indeed, when $\mathcal{J} = \mathcal{K} = [\![1, n]\!]$, we have definition 3.2. Based on this observation, it can easily be shown that the properties of octagons hold for partial octagons, and hence the set of partial octagons can be the base set for an abstract domain.

The choice of rotated basis to generate for a partial octagon depends on the problem to solve. Different octagonalization heuristics are presented in section 4.3.2.

## 3.5. Conclusion

In this chapter, we have presented two equivalent and computer representable forms for the octagons. We have shown that the set of closed octagons by the modified version of the Floyd–Warshall algorithm forms a complete lattice. Moreover, we have defined the octagonal splitting operator and size function as well as the Galois connection between the octagons and the boxes with floating-point bounds. With all these elements, we were able to define the octagons as an abstract domain in CP. Since the octagon abstract domain relies on the generation of all the possible basis, we defined the partial octagon set. This set is the base set of an abstract domain. The next chapter gives additional details regarding the implementation of a solving method based on octagons.

# 4

# Octagonal Solving

In this chapter we define a consistency based on octagons as well as a propagation schema. Due to this consistency and to the octagonal abstract domain defined in the previous chapter, we obtain a resolution method based on octagons. A prototype of this method has been implemented in a continuous solver. The details of this implementation are given in this chapter along with some preliminary results.

Consider a constraint satisfaction problem (CSP) on variables $(v_1 \dots v_n)$ in $\mathbb{R}^n$. The first step is to represent this problem in an octagonal form. We detail here the construction of an octagonal CSP from a CSP. We show that these two systems are equivalent.

## 4.1. Octagonal CSP

First, the CSP is associated with an octagon, by stating all the possible non-redundant octagonal constraints $\pm v_i \pm v_j \leq c$ for $i, j \in [\![1, n]\!]$. The constants $c$ represent the bounds of the boxes in the rotated basis, or octagonal boxes, and the bound of the boxes in the canonical basis. These constants are dynamically modified during the solving process. They are initialized to $+\infty$.

The rotations defined in the previous chapter (definition 3.5) introduce new axes, corresponding to new variables $v_i^{i,j}$ and $v_j^{i,j}$ in the $(i, j)$-rotated basis. These variables are redundant with the variables $v_i$

and $v_j$ in the canonical basis. If variables $v_i$ and $v_j$ are linked through constraints, then the new variables $v_i^{i,j}$ and $v_j^{i,j}$ have to be linked as well. The CSP constraints $C_1 \dots C_p$ have to be rotated as well. We explain here how constraints are rotated.

Given a function $f$ on variables $(v_1 \dots v_n)$ in the canonical basis $B$, the expression of $f$ in the $(i, j)$-rotated basis is obtained by symbolically replacing the $i$-th and $j$-th coordinates by their expressions in $B_\alpha^{i,j}$. Precisely, variable $v_i$ is replaced by $(\cos(\alpha)v_i^{i,j} - \sin(\alpha)v_j^{i,j})$ and variable $v_j$ by $(\sin(\alpha)v_i^{i,j} + \cos(\alpha)v_j^{i,j})$ where $v_i^{i,j}$ and $v_j^{i,j}$ are the coordinates for $v_i$ and $v_j$ in $B_\alpha^{i,j}$. The other variables are unchanged.

DEFINITION 4.1 (Rotated constraint).– *Given a constraint $C$ holding on variables $(v_1 \dots v_n)$. The $(i, j)$-*rotated constraint* $C^{i,j}$ is the constraint obtained by replacing each occurrence of $v_i$ by $(\cos(\alpha)v_i^{i,j} - \sin(\alpha)v_j^{i,j})$ and each occurrence of $v_j$ by $(\sin(\alpha)v_i^{i,j} + \cos(\alpha)v_j^{i,j})$.*

EXAMPLE 4.1 (Rotated constraint).– Let $v_1$ and $v_2$ be two variables. Let $C$ be the constraint $2v_1 + v_2 \leq 3$. The $(1, 2)$-rotated constraint is:

$$C^{1,2} \equiv 2\left(\cos\left(\frac{\pi}{4}\right)v_1^{1,2} - \sin\left(\frac{\pi}{4}\right)v_2^{1,2}\right) + \left(\sin\left(\frac{\pi}{4}\right)v_1^{1,2} + \cos\left(\frac{\pi}{4}\right)v_2^{1,2}\right) \leq 3$$

given that $\sin\left(\frac{\pi}{4}\right) = \cos\left(\frac{\pi}{4}\right) = \frac{1}{\sqrt{2}}$, it can be simplified as:

$$C^{1,2} \equiv 3v_1^{1,2} - v_2^{1,2} \leq 3\sqrt{2}$$

Figure 4.1 graphically compares the constraints in the canonical basis to the rotated basis. Figure 4.1(a) shows the two constraints given in the initial CSP. The equivalent constraints after rotation of $\frac{\pi}{4}$ are given in Figure 4.1(b). Rotated constraints live in the $(1, 2)$-rotated basis $B^{1,2}$.

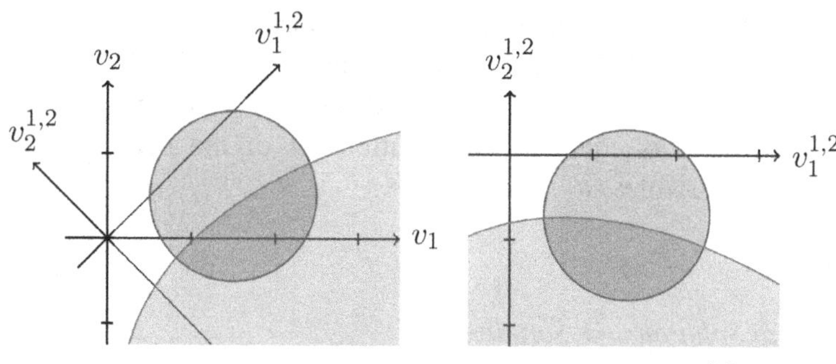


a) The constraints in the canonical basis. b) The constraints in $B^{1,2}$ the $(1,2)$-rotated basis.

**Figure 4.1.** *Example of rotated constraints: comparison between the initial CSP a) and the rotated CSP b). For a color version of the figure, see www.iste.co.uk/pelleau/abstractdomains.zip*

Given a continuous CSP on variables $(v_1 \ldots v_n)$, with continuous domains $(\hat{D}_1 \ldots \hat{D}_n)$ and constraints $(C_1 \ldots C_p)$, we defined an octagonal CSP by adding the rotated variables, the rotated constraints and the rotated domains stored in a difference bound matrix.

To sum up, an octagonal CSP contains:

– the initial variables $(v_1 \ldots v_n)$;

– the rotated variables $(v_1^{1,2}, v_2^{1,2}, v_1^{1,3}, v_3^{1,3} \ldots v_n^{n-1,n})$, where $v_i^{i,j}$ is the $i$-th variable in the $(i,j)$-rotated basis $B_\alpha^{i,j}$;

– the initial constraints $(C_1 \ldots C_p)$;

– the rotated constraints $(C_1^{1,2}, C_1^{1,3} \ldots C_1^{n-1,n} \ldots C_p^{n-1,n})$;

– the initial domains $(\hat{D}_1 \ldots \hat{D}_n)$;

– a difference bound matrix which represents the rotated domains. It is initialized with the bounds of the regular domains $\hat{D}_i$ for the cells at position $(2i, 2i-1)$ and $(2i-1, 2i)$ for $i \in [\![1, n]\!]$, and $+\infty$ everywhere else.

In these conditions, the initial CSP is equivalent to this octagonal CSP, restricted to the variables $v_1 \ldots v_n$, as shown in the following proposition.

PROPOSITION 4.1.– Consider a CSP on variables $(v_1 \ldots v_n)$, with domains $(\hat{D}_1 \ldots \hat{D}_n)$ and constraints $(C_1 \ldots C_p)$, and the corresponding octagonal CSP as defined above. The solution set of the original CSP $S$ is equal to the solution set of the octagonal CSP restricted to variables $(v_1 \ldots v_n)$.

PROOF 4.1.–

*Octagonal Solution $\Longrightarrow$ Solution*

First we prove that an octagonal solution is a solution in the original CSP. Let $s \in \mathbb{R}^n$ a solution of the octagonal CSP restricted to variables $(v_1 \ldots v_n)$. We have $C_1(s) \ldots C_p(s)$ and $s \in \hat{D}_1 \times \cdots \times \hat{D}_n$. Hence $s$ is a solution for the original CSP. $\diamond$

*Solution $\Longrightarrow$ Octagonal Solution*

We now prove that a solution for the original CSP is an octagonal solution. Reciprocally, let $s \in \mathbb{R}^n$, a solution of the original CSP. The initial constraints $(C_1 \ldots C_p)$ are true for $s$. Let us show that there exist values for the rotated variables such that the rotated constraints are true for $s$. Let $i, j \in [\![1, n]\!]$, $i \neq j$ and $k \in [\![1, p]\!]$. Let $C_k$ be a constraint in the original CSP and $C_k^{i,j}$ the corresponding rotated constraint. From definition 4.1 $C_k^{i,j}(v_1 \ldots v_{i-1}, \cos(\alpha)v_i^{i,j} - \sin(\alpha)v_j^{i,j}, v_{i+1} \ldots \sin(\alpha)v_i^{i,j} + \cos(\alpha)v_j^{i,j} \ldots v_n) \equiv C_k(v_1 \ldots v_n)$. Let us define the two reals $s_i^{i,j} = \cos(\alpha)s_i + \sin(\alpha)s_j$ and $s_j^{i,j} = -\sin(\alpha)s_i + \cos(\alpha)s_j$ the image of $s_i$ and $s_j$, by the rotation of angle $\alpha$. By reversing the rotation, $\cos(\alpha)s_i^{i,j} - \sin(\alpha)s_j^{i,j} = s_i$ and $\sin(\alpha)s_i^{i,j} + \cos(\alpha)s_j^{i,j} = s_j$, thus $C_k^{i,j}(s_1 \ldots s_i^{i,j}, \ldots s_j^{i,j} \ldots s_n) = C_k(s_1 \ldots s_n)$ is true. It remains to check that $(s_1 \ldots s_i^{i,j}, \ldots s_j^{i,j} \ldots s_n)$ is in the rotated domain, which is true because the difference bound matrix is initialized at $+\infty$. Hence $s$ is a solution for the octagonal CSP. $\square$

For a CSP on $n$ variables, this representation has an order of magnitude $n^2$. Indeed, the octagonal CSP has $n^2$ variables and domains, and at most $p\left(\frac{n(n-1)}{2} + 1\right)$ constraints. Of course, many of these objects are redundant. We explain in the next sections how to use this redundancy to speed up the solving process.

As mentioned, when defining abstract domains in CP section 2.2.3, propagators have to be *ad hoc* as they depend both on the abstract domain and constraints. The next sections show how the consistency and propagation scheme can be defined for the octagons.

## 4.2. Octagonal consistency and propagation

First, we generalize the definition of Hull-consistency on intervals (definition 1.17) to the octagons and define the propagators for the rotated constraints. Then we define an efficient propagation scheme for both the rotated constraints and octagonal constraints using the modified version of Floyd-Warshall algorithm (algorithm 3.1).

### 4.2.1. *Octogonal consistency*

We generalize the definition of Hull-consistency on intervals for any continuous constraint to the octagons. With the intersection of boxes representation, we show that any propagator for Hull-consistency on boxes can be extended to a propagator on the octagons. For a given $n$-ary relation on $\mathbb{R}^n$, there is a unique smallest octagon (w.r.t. inclusion) which contains the solutions of this relation, as shown in the following proposition.

PROPOSITION 4.2.– Consider a constraint $C$ (respectively a constraints sequence $(C_1 \dots C_p)$), and $S_C$ its set of solutions (respectively $S$). Then there exist a unique octagon $\mathbf{O}$ such that: $S_C \subseteq \mathbf{O}$ (respectively $S \subseteq \mathbf{O}$), and for all octagons $\mathbf{O}'$, $S_C \subseteq \mathbf{O}'$ implies $\mathbf{O} \subseteq \mathbf{O}'$. $\mathbf{O}$ is the unique smallest octagon containing the solutions, w.r.t. inclusion. $\mathbf{O}$ is said Oct-consistent for the constraint $C$ (respectively constraints sequence $(C_1 \dots C_p)$).

PROOF 4.2.– Octagons are closed by intersection (remark 3.4) and the set of floating bound octagons forms a finite lattice and thus complete (proposition 3.3). Hence the proof comes directly from proposition 2.5. □

PROPOSITION 4.3.– Let $C$ be a constraint, and $C^{i,j}$ the $(i, j)$-rotated constraint for $i, j \in [\![1, n]\!]$. Let $\mathbf{B}^{i,j}$ be the Hull-consistent box for $C^{i,j}$, and $\mathbf{B}$, the Hull-consistent box for $C$. The Oct-consistent octagon for $C$ is the intersection of all the $\mathbf{B}^{i,j}$ and $\mathbf{B}$.

PROOF 4.3.– The intersection of boxes representation is an exact representation for octagons (proposition 3.1). The proof comes directly from proposition 4.2. □

The consistency indicates how to filter the inconsistent values for a constraints, We now have to chose in which order the constraints are propagated.

### 4.2.2. *Propagation scheme*

The modified version of Floyd-Warshall shortest path (algorithm 3.1) efficiently and optimally propagates octagonal constraints [DEC 89]. The initial constraints $(C_1 \dots C_p)$ and the rotated constraints $(C_1^{1,2} \dots C_p^{n-1,n})$ are filtered using the Oct-consistency. So we need to integrate both propagators in the same propagation scheme. For octagonal constraints corresponding to the bounds of the boxes, we rely on information gathered during the propagation of these constraints to propagate the initial and rotated constraints. This allows us to take full advantage of relational properties of octagons. It is important to note that all the propagators defined above are monotonous and complete; we can then combine them in any order to compute the consistency, as shown in proposition 2.5 or in [BEN 96].

The key idea for the propagation scheme is to interleave the refined Floyd-Warshall algorithm and the propagators for the initial and rotated constraints. A pseudocode is given in algorithm 4.1. During the first iteration of the propagation loop, all the propagators for the initial and rotated constraints are executed. This first stage reduces the domains with respect to the constraints of the problem. As the domains are stored in the difference bound matrix, the values in the matrix are thus modified and the modified version of Floyd-Warshall must be applied to compute the new minimal bounds. The second part of the propagation loop corresponds to algorithm 3.1, except that every time an element is modified, the corresponding propagators are added to the set of propagators to be executed. The propagators in this set are applied, once the propagation of the octagonal constraints is done. Hence, if an element is modified several times, its corresponding propagators are called only once. We can say that this propagation scheme is guided by the additional information of the relational domain.

An example of an iteration of the propagation loop is given in Figure 4.2: propagators of the initial constraints $\rho_{C_1}, \rho_{C_2}$ are first applied (Figure 4.2(a)), then the propagators of the rotated constraints are applied $\rho_{C_1^{1,2}}, \rho_{C_2^{1,2}}$ (Figure 4.2(b)). The two boxes are made consistent with respect to each other using the refined Floyd-Warshall algorithm (Figure 4.2(c)). In this example, the box in the canonical basis is slightly reduced, thus the propagators $\rho_{C_1}, \rho_{C_2}$ have to be applied again. The application of these propagators do not modify the box in the canonical basis and the consistent octagon corresponds to the intersection of theses two boxes (Figure 4.2(d)).

We show here that the propagation as defined in algorithm 4.1 is correct; in other words, it computes the consistent octagon for a sequence of constraints.

PROPOSITION 4.4 (Correctness).– Given a CSP on variables $(v_1 \ldots v_n)$, domains $(\hat{D}_1 \ldots \hat{D}_n)$, and constraints $(C_1 \ldots C_p)$. Suppose that each constraint $C$ comes with a propagator $\rho_C$ such that $\rho_C$ reaches the Hull-consistency, that is $\rho_C(\hat{D}_1 \times \cdots \times \hat{D}_n)$ returns a hull-consistent box for $C$. Then the propagation scheme, as defined in algorithm 4.1, computes the Oct-consistent octagon for $C_1 \ldots C_p$.

PROOF 4.4.– This proposition derives from proposition 4.3, and the propagation scheme in algorithm 4.1. The propagation scheme is defined so as to stop when propagSet is empty. This happens when the difference bound matrix is no more modified. Indeed, the propagators are added to propagSet only when an element in the difference bound matrix is modified. Thus when the propagation scheme terminates, the octagonal constraints are consistent. Moreover, as the elements in the difference bound matrix are not changed, it follows that the propagators did not modify the variables domain. Hence, the final octagon is stable by application of all $\rho_{C_k^{i,j}}$, for all $k \in [\![1,p]\!]$ and $i,j \in [\![1,n]\!]$. By hypothesis, the propagators reach consistency, thus the boxes are hull-consistent for all the initial and rotated constraints. From proposition 4.3, the returned octagon is Oct-consistent. □

**Algorithm 4.1.** *Pseudo code for the propagation loop mixing the modified version of Floyd Warshall algorithm (octagonal constraints propagation) and the initial and rotated propagators $\rho_{C_1} \dots \rho_{C_p}, \rho_{C_1^{1,2}} \dots \rho_{C_p^{n-1,n}}$, for an octagonal CSP as defined section 4.1. For a color version of the algorithm, see www.iste.co.uk/pelleau/abstractdomains.zip*

```
float dbm[2n][2n]        /* difference bound matrix containing the variables domain */
set of propagators propagSet                         /* set of propagators to execute */
int i, j, k, i', j'                        /* indexes for the difference bound matrix */
float m                                                        /* auxiliary variable */

propagSet ← {ρ_{C_1}, ..., ρ_{C_p}, ρ_{C_1^{1,2}}, ..., ρ_{C_p^{n-1,n}}}
while propagSet ≠ ∅ do
                                       /* initial and rotated constraints propagation */
  apply all the propagators in propagSet
  propagSet ← ∅
                                                /* octagonal constraints propagation */
  for k from 1 to n do
    for i from 1 to 2n do
      for j from 1 to 2n do
        m ← min( dbm[i][2k] + dbm[2k][j], dbm[i][2k − 1] + dbm[2k − 1][j],
                 dbm[i][2k − 1] + dbm[2k − 1][2k] + dbm[2k][j],
                 dbm[i][2k] + dbm[2k][2k − 1] + dbm[2k − 1][j]              )
        if m < dbm[i][j] then
          dbm[i][j] ← m                                       /* update of the DBM */
          propagSet ← propagSet ∪{ρ_{C_1^{i,j}} ... ρ_{C_p^{i,j}}}   /* get the propagators to
          apply */
        end if
      end for
    end for
  end for
  for i from 1 to 2n do
    for j from 1 to 2n do
      i' ← (i mod 2 = 0) ? i − 1 : i + 1
      j' ← (j mod 2 = 0) ? j − 1 : j + 1
      if (dbm[i][i'] + dbm[j'][j])) < dbm[i][j] then
        dbm[i][j] ← dbm[i][i'] + dbm[j'][j]
        propagSet ← propagSet ∪{ρ_{C_1^{i,j}} ... ρ_{C_p^{i,j}}}
      end if
    end for
  end for
  for i from 1 to 2n do
    if dbm[i][i] < 0 then
      return failure
    else
      dbm[i][i] ← 0
    end if
  end for
end while
```

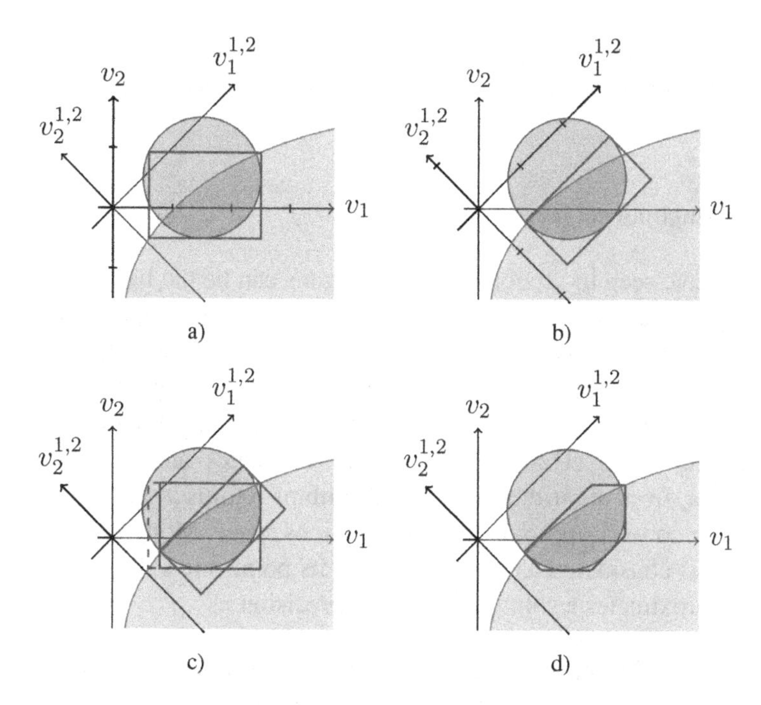


**Figure 4.2.** *Example of an iteration of the propagation for the Oct-consistency: an usual consistency algorithm is applied in each basis a) and b) then the different boxes are made consistent using the modified Floyd-Warshall algorithm c). The consistent octagon is given d)*

The refined Floyd-Warshall has a time complexity of $O(n^3)$. For each round in its loop, in the worst case we add all the propagators to the set of propagators to execute, that is in the worst case $p\left(\frac{n(n-1)}{2}+1\right)$ propagators. Thus the time complexity for the propagation scheme in algorithm 4.1 is $O(n^3 + pn^2)$. In the end, the octagonal propagation uses both representations of octagons. It takes advantage both of the relational property of the octagonal constraints (Floyd-Warshall), and of the usual constraint propagation on boxes (propagators). This adds to the cost of computing the octagon, but is expected to give a better precision in the end.

Now that we have defined all the necessary components for the solving process for the octagons, namely the octagonal consistency, the octagonal splitting operator, the octagonal precision and the octagons abstract domain, we can go a step further and define a fully octagonal solver.

## 4.3. Octagonal solver

We have seen in section 3.4 that octagons can be the base set for an abstract domain and that with the consistency and the propagation loop described previously, we obtain an octagonal solver. The solving process is the same as that in algorithm 2.1 using octagons. Octagons being closed under intersection (H1) and $\mathbb{O}$ not having infinite decreasing chain (H2), this algorithm terminates and is complete. Moreover, from algorithm 2.1, this algorithms returns a set of octagons which union over-approximates the solutions space. More precisely, an octagon is considered as a solution if all its points are solutions or if it over-approximates a solution set with a precision $r$.

To guide the search space exploration, we can define heuristics for the choice of the variable. Several heuristics are presented in the next section.

### 4.3.1. *Variables heuristics*

An important feature of a solver is the heuristic used to choose the variable. It chooses which variable to split. For continuous constraints, the variable with the largest domain size is chosen to be split in two. By splitting the largest domain, this heuristic can arrive faster to the precision. The three following heuristics derived from this strategy and differ on the set of variables that can be chosen. Let $V'$ be the set of all the variables in an octagonal CSP; $V' = (v_1 \dots v_n, v_1^{1,2}, v_2^{1,2} \dots v_{n-1}^{n-1,n}, v_n^{n-1,n})$ which is re-numbered $V' = (v_1 \dots v_n, v_{n+1} \dots v_{n^2})$ for more simplicity.

DEFINITION 4.2 (Largest-First heuristic (LF)).– *Let* $V' = (v_1 \ldots v_n, v_{n+1} \ldots v_{n^2})$ *be the set of all the variables in the octagonal CSP. The variable to split is the variable* $v_i$ *which realizes the maximum of*

$$\arg\max_{i \in [\![1,n^2]\!]} \left(\overline{D_i} - \underline{D_i}\right)$$

As for the usual continuous split, the variable to split is chosen among all the variables.

DEFINITION 4.3 (Largest-Can-First heuristic (LCF)).– *Let* $V' = (v_1 \ldots v_n, v_{n+1} \ldots v_{n^2})$ *be the set of all the variables in the octagonal CSP. The variable to split is the variable* $v_i$ *realizing the maximum of*

$$\arg\max_{i \in [\![1,n]\!]} \left(\overline{D_i} - \underline{D_i}\right)$$

In this strategy, the choice is restricted to the initial variables which are the variables in the original CSP.

DEFINITION 4.4 (Largest-Oct-First heuristic (LOF)).– *Let* $V' = (v_1 \ldots v_n, v_{n+1} \ldots v_{n^2})$ *be the set of all the variables in the octagonal CSP. The variable to split is the variable* $v_i$ *which realizes the maximum of*

$$\arg\max_{i \in [\![n+1,n^2]\!]} \left(\overline{D_i} - \underline{D_i}\right)$$

In this strategy, the choice is restricted to the rotated variables which are the variables generated by the rotations.

The three choice heuristics presented above do not take into account the correlation between the variables. Moreover, the variable which has the largest domain can be of a basis that is of little interest for the problem, or one whose domains have a wide range because the constraints are poorly propagated in this basis. We thus define an octagonal strategy.

DEFINITION 4.5 (Oct-Split heuristic (OS)).– *This strategy relies on the same remark as for definition 3.7: the variable to split is the variable* $v_k^{i,j}$ *which realizes the maximum of*

$$\min_{i,j \in [\![1,n]\!]} \left( \max_{k \in [\![1,n]\!]} \left( \overline{D_k^{i,j}} - \underline{D_k^{i,j}} \right) \right)$$

The strategy is as follows: first choose a promising basis, that is, a basis in which the boxes are tight (choose $i, j$). Then take the worst variable in this basis as usual (choose $k$). This heuristic aims at reaching the octagonal precision faster (definition 3.7).

Figure 4.3 shows examples of the different variables heuristics. In the two examples, we consider the octagon obtained by the intersection of two boxes. In the first example, strategies Largest-First and Largest-Can-First choose to split domain $D_1$ (along the dashed line). While strategies Largest-Oct-First and Oct-Split choose to split domain $D_{1,2}$ (along the dotted line). In the second example, strategies Largest-First and Largest-Oct-First choose to split domain $D_{1,2}$ (along the dashed line). While strategies Largest-Can-First and Oct-Split choose to split domain $D_1$ (along the dotted line).

For these two examples, the split along the dotted line seems more relevant. In the first example, splitting along the dotted line reduces the size of the domains in the rotated basis which is interesting for this octagon as it corresponds to the box in the rotated basis. Similarly, in the second example, reducing the size of the domains in the canonical basis is more relevant as the octagon corresponds to the box in the canonical basis.

#### 4.3.2. *Octogonalization heuristic*

Octagonalization heuristics create partial octagons. For the octagons, all the $n(n+1)/2$ possible basis are generated, which for problems of high dimension (large value for $n$) created an octagon with a large number of faces. As previously mentioned, maybe some of the generated basis are less interesting than others with respect to the problem to solve. Thus, we define different strategies to determine a subset of bases to generate. Given a problem, we try to find, by symbolically looking at the constraints, which bases are more likely to improve the search. In other words, we try to determine in which bases the problem is easier to solve by only looking at the constraints. This step is done once at the creation of the octagon unlike in [GOL 10] where the basis changes after each call to the splitting operator in the solving process.

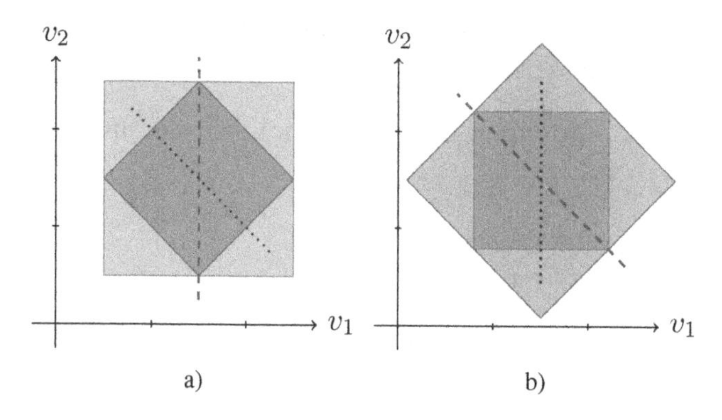


**Figure 4.3.** *Examples of the different variable heuristics. Figure 4.3(a), the LF and LCF strategies split along the red dashed line while the LOF and OS strategies split along the blue dotted line. Figure 4.3(b), the LF and LOF strategies split along the red dashed line while the LCF and OS strategies split along the blue dotted line. For a color version of the figure, see www.iste.co.uk/pelleau/abstractdomains.zip*

DEFINITION 4.6 (Constraint-Based heuristic (CB)).– *Consider a CSP on variables $(v_1 \ldots v_n)$ to octogonalize. The basis $B_\alpha^{i,j}$ is generated only if the variables $v_i$ and $v_j$ appear within the same constraint.*

If $v_i$ and $v_j$ appear in the constraint $C$, then $C$ links $v_i$ with $v_j$. We say that there exists a relation between $v_i$ and $v_j$. The idea behind this heuristic is to emphasize a relation that already exists between two variables. In the worst case, all the pairs of variables appear within a constraint, thus, all the bases will be generated.

DEFINITION 4.7 (Random heuristic (R)).– *Consider a CSP on variables $(v_1 \ldots v_n)$ to octogonalize. Among all the possible bases generates* one *random basis. Choosing $i$ and $j$ with $i < j$ then generates the basis $B_\alpha^{i,j}$.*

We set the number of generated bases to one which is purely arbitrary and can be random for a pure random heuristic.

DEFINITION 4.8 (Strongest-Link heuristic (SL)).– *Consider a CSP on variables $(v_1 \dots v_n)$ to octogonalize. This heuristic generates only one basis, the one for which the variables have the strongest link. In other words, the basis corresponding to the pair of variables which appears in most constraints.*

EXAMPLE 4.2 (Strongest-Link heuristic).– Given the following constraints:

$$\begin{cases} v_1 + v_2 + v_1 \times v_2 \leq 3 \\ \cos(v_1) + v_3 \leq 10 \\ v_1 \times v_3 \geq 1 \end{cases}$$

The basis $B_\alpha^{1,3}$ is generated as variables $v_1$ and $v_3$ appear together in two constraints, while the pair $\{v_1, v_2\}$ appear in only one constraint.

The idea of the Strongest-Link heuristic is to emphasize the strongest relation between two variables. We want the new variables to be as constrained as possible so that the solving process will have more chances of reducing their domains.

The last heuristic rely on the notion of promising scheme.

DEFINITION 4.9 (Promising scheme).– *Let $(v_1 \dots v_n)$ be the variables set. We call* promising scheme *expressions corresponding to the following patterns: $\pm v_i \pm v_j$ or $\pm v_i \times v_j$ for $i, , j \in [\![1, n]\!]$.*

These schemas are said to be promising as they can easily simplified in the corresponding rotated basis.

EXAMPLE 4.3 (Promising scheme).– Let $v_1$ and $v_2$ be two variables. Consider the expression $v_1 + v_2$, the corresponding expression in the $(1, 2)$-rotated basis is:

$$\begin{aligned} v_1 + v_2 &= av_1^{1,2} - av_2^{1,2} + av_1^{1,2} + av_2^{1,2} \\ &= 2av_1^{1,2} \end{aligned}$$

where $a = \cos\left(\frac{\pi}{4}\right) = \sin\left(\frac{\pi}{4}\right)$. Similarly, consider the expression $v_1 \times v_2$. The corresponding expression in the $(1, 2)$-rotated basis is:

$$\begin{aligned} v_1 \times v_2 &= \left(av_1^{1,2} - av_2^{1,2}\right) \times \left(av_1^{1,2} + av_2^{1,2}\right) \\ &= \left(av_1^{1,2}\right)^2 - \left(av_2^{1,2}\right)^2 \end{aligned}$$

DEFINITION 4.10 (Promising heuristic (P)).– *Consider a CSP on variables* $(v_1 \ldots v_n)$ *to octogonalize. This heuristic generates the basis* $B_\alpha^{i,j}$*, maximizing the number of promising patterns in the constraints.*

EXAMPLE 4.4 (Promising heuristic).– Given the following constraints:

$$\begin{cases} v_1 + v_2 + v_1 \times v_2 \leq \ 3 \\ \cos(v_1) + v_3 \leq 10 \\ v_1 \times v_3 \geq \ 1 \end{cases}$$

The basis $B_\alpha^{1,2}$ is generated as there are two promising patterns with variables $v_1$ and $v_2$ ($v_1 + v_2$ and $v_1 \times v_2$) while for the pair $\{v_1, v_3\}$ there is only one promising pattern ($v_1 \times v_3$).

With this heuristic, we want to reduce the number of multiple occurrences of the new variables in the rotated constraints. We know that implementations of the Hull-consistency is sensitive to multiple occurrences of variables. By reducing the number of multiple occurrences, we hope to have a more efficient consistency and thus, to have a faster solving process.

## 4.4. Experimental results

This section compares the results obtained with the octagonal solver to those obtained with a traditional interval solver on classical benchmarks.

### 4.4.1. *Implementation*

We have implemented a prototype of the octagonal solver, with Ibex, a C++ library for continuous constraints [CHA 09a]. We use the Ibex implementation of *HC4-Revise* [BEN 99] to propagate the constraints. The octagons are implemented with their matrix representation (DBM). Additional rotated variables and constraints are posted and dealt with as explained in the previous section.

An important point is the rotation of the constraints. The HC4 algorithm is sensitive to multiple occurrences of the variables, and the symbolic rewriting defined in section 4.1 creates multiple occurrences. Thus, the HC4 propagation on the rotated constraints could be very poor if performed directly on the rotated constraints. It is necessary to simplify the rotated constraints with respect to the number of multiple occurrences for the variables. We use the `simplify` function of Mathematica to do this. The computation time indicated below does not include the time for this treatment, however, it is negligible compared to the solving times.

The propagator is an input of our method: we used a standard one (HC4). Further work include adding propagators such as [ARA 10], which better takes into account the symbolic expression of the constraint to improve the propagation.

### 4.4.2. *Methodology*

We have tested the prototype octagonal solver on problems from the Coconut benchmark[1]. These problems have been chosen depending on the type of the constraints (inequations, equations, or both) and on the number of variables.

First, we have compared the computation time to find the first solution or all the solutions with the intervals and the octagons. For these experiments, we have fixed the precision parameter $r$ to $0.01$. For

1 This benchmark can be found at www.mat.univie.ac.at/neum/glopt/coconut/.

all the other experiments, the precision parameter $r$ is set to 0.001. The second experiment concerns the study of the different heuristics for the choice of variable presented in section 4.3.1. The last series of tests, compares the results obtained by the partial octagons generated by the different octagonalization heuristics (section 4.3.2) to the results obtained with octagons and intervals.

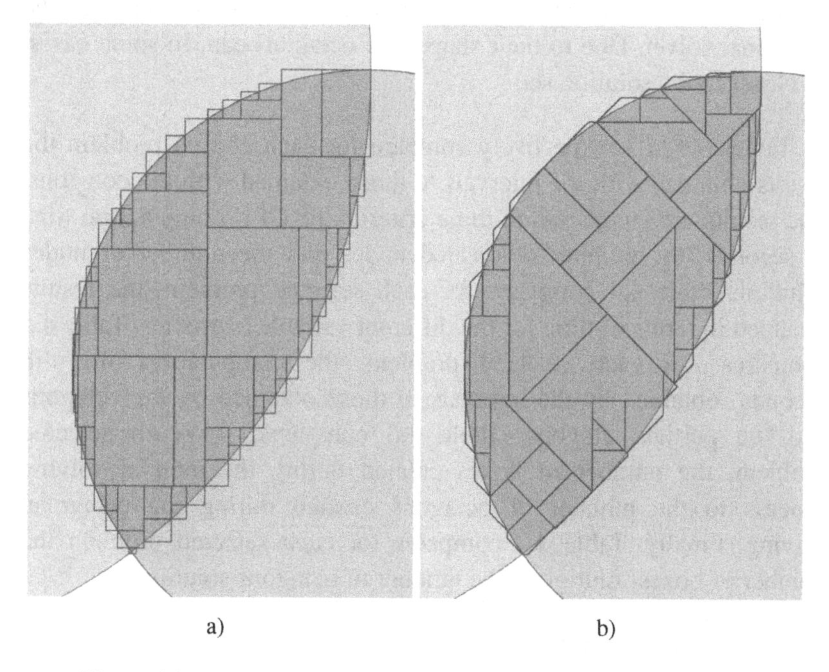

a) b)

**Figure 4.4.** *Comparison of the results obtained with the usual interval resolution a) to the results obtained with the octagonal resolution b), given an intersection problem and a time limit of 10 ms*

All experiments were performed with Ibex 1.18 on a MacBook Pro Intel Core 2 Duo 2.53 GHz. All were performed with the same configurations in IBEX. In particular, the same propagators were used in order to more accurately compare the results obtained with the octagonal resolution to those obtained with the intervals. In addition, we have set a time limit to 3 h in all cases.

### 4.4.3. *Results*

Figure 4.4 compares the results obtained with the interval resolution (Figure 4.4(a)) to the results obtained with the octagonal solver (Figure 4.4(b)) given a time limit of 10 ms. We can see that with the standard interval solver, there is a staircase effect caused by the boxes on the limit of the solution set. This effect is less present with the octagonal solver. Due to their shape, the octagons can, in some cases, be closer to the solution set.

Tables 4.1–4.3 respectively compare for each chosen problem the results obtained with the intervals to those obtained with the octagons. The results are compared on three criteria, the CPU computation time in seconds, the number of created nodes and the number of nodes solution. Table 4.4 compares for each selected problem, the results obtained in terms of time for the different variable heuristics. Table 4.5 compares, for each selected problem, the computation time (in seconds) obtained by the intervals to those obtained by the octagons and the partial octagons. Table 4.6 compares, for each selected problem, the number of boxes created during the interval solving process to the number of octagons created during the octagonal solving. Finally, Table 4.7 compares for each selected problem the number of boxes solution to the number of octagons solution.

In all the tables, the first three columns give the name of the problem, the number of variable and the type of the constraints. Also, the dash symbol '-' stands for 'time out' (3 h).

### 4.4.4. *Analysis*

We analyse here the results obtained with the different experiments.

#### 4.4.4.1. *Comparison one solution versus all the solutions*

Table 4.1 compares the results in term of CPU time needed to find the first solution or all the solutions of a problem using either the intervals or the octagons. Finding the first solution is often faster using the octagons than with the intervals. Indeed, the octagons being more precise, they are usually closer to the solutions than the boxes, and therefore, finding the first solution is faster. However, finding all the solutions can be longer with the octagons than with the boxes. A first explanation is that constraints in problems `brent-10`, `pramanik` and `trigo1` contain many multiple occurrences of variable, which highly increases the number of multiple occurrences of rotated variables, despite the call to the `simplify` function in Mathematica.

| | | | First solution | | All solutions | |
|---|---|---|---|---|---|---|
| Name | # var | ctr type | $\mathbb{B}$ | $\mathbb{O}$ | $\mathbb{B}$ | $\mathbb{O}$ |
| h75 | 5 | $\leq$ | 41.40 | **0.03** | - | - |
| hs64 | 3 | $\leq$ | **0.01** | 0.05 | - | - |
| h84 | 5 | $\leq$ | 5.47 | **2.54** | - | **7238.74** |
| KinematicPair | 2 | $\leq$ | **0.00** | **0.00** | 53.09 | **16.56** |
| pramanik | 3 | $=$ | 28.84 | **0.16** | **193.14** | 543.46 |
| trigo1 | 10 | $=$ | 18.93 | **1.38** | **20.27** | 28.84 |
| brent-10 | 10 | $=$ | 6.96 | **0.54** | **17.72** | 105.02 |
| h74 | 4 | $=$ $\leq$ | 305.98 | **13.70** | 1304.23 | **566.31** |
| fredtest | 6 | $=$ $\leq$ | 3146.44 | **19.33** | - | - |

**Table 4.1.** *Results on problems from the Coconut benchmark. For each problem, we give the CPU time in second to find the first solution or all the solutions with the intervals ($\mathbb{B}$) and the octagons ($\mathbb{O}$)*

| | | | First solution | | All solutions | |
|---|---|---|---|---|---|---|
| Name | # var | ctr type | $\mathbb{B}$ | $\mathbb{O}$ | $\mathbb{B}$ | $\mathbb{O}$ |
| h75 | 5 | $\leq$ | 1 024 085 | **149** | - | - |
| hs64 | 3 | $\leq$ | 217 | **67** | - | - |
| h84 | 5 | $\leq$ | 87 061 | **1 407** | - | **22 066 421** |
| KinematicPair | 2 | $\leq$ | 45 | **23** | 893 083 | **79 125** |
| pramanik | 3 | $=$ | 321 497 | **457** | 2 112 801 | **1 551 157** |
| trigo1 | 10 | $=$ | 10 667 | **397** | 11 137 | **5 643** |
| brent-10 | 10 | $=$ | 115 949 | **157** | 238 777 | **100 049** |
| h74 | 4 | $=$ $\leq$ | 8 069 309 | **138 683** | 20 061 357 | **1 926 455** |
| fredtest | 6 | $=$ $\leq$ | 29 206 815 | **3 281** | - | - |

**Table 4.2.** *Results on problems from the Coconut benchmark. For each problem, we give the number of created nodes to find the first solution or all the solutions with the intervals ($\mathbb{B}$) and the octagons ($\mathbb{O}$)*

| Name | # var | ctr type | $\mathbb{B}$ | $\mathbb{O}$ |
|---|---|---|---|---|
| h75 | 5 | $\leq$ | - | - |
| hs64 | 3 | $\leq$ | - | - |
| h84 | 5 | $\leq$ | - | **10 214 322** |
| KinematicPair | 2 | $\leq$ | 424 548 | **39 555** |
| pramanik | 3 | $=$ | **145 663** | 210 371 |
| trigo1 | 10 | $=$ | **12** | 347 |
| brent-10 | 10 | $=$ | 854 | **142** |
| h74 | 4 | $=$ $\leq$ | 700 669 | **183 510** |
| fredtest | 6 | $=$ $\leq$ | - | - |

**Table 4.3.** *Results on problems from the Coconut benchmark. For each problem, we give the number of solution nodes to find the first solution or all the solutions with the intervals ($\mathbb{B}$) and the octagons ($\mathbb{O}$)*

| Name | # var | ctr type | LF | LOF | LCF | OS |
|---|---|---|---|---|---|---|
| brent-10 | 10 | $=$ | - | - | 362.77 | **337.48** |
| o32 | 5 | $\leq$ | 104.62 | 122.82 | **40.60** | 40.74 |
| ipp | 8 | $=$ | 5 105.17 | 5 787.5 | 282.319 | **279.36** |
| trigo1 | 10 | $=$ | - | - | 256.71 | **253.53** |
| KinematicPair | 2 | $\leq$ | **59.72** | 60.74 | 62.91 | 60.78 |
| nbody5.1 | 6 | $=$ | 105.93 | 121.33 | 27.33 | **27.08** |
| pramanik | 3 | $=$ | 396.23 | 414.76 | **240.93** | 240.96 |
| h74 | 4 | $=$ $\leq$ | 2896.58 | 4036.89 | 1553.67 | **647.76** |

**Table 4.4.** *Comparison of the different variable heuristics on problems from the Coconut benchmark. For each problem, we give the CPU time in seconds to solve it using the Largest-First heuristic (LF), the Largest-Oct-First (LOF), the Largest-Can-First (LCF) and the Oct-Split (OS)*

Looking now to the number of created nodes during the resolution (Table 4.2), the number of octagons created is often smaller than the number of boxes created during the resolution. This comes from the fact that octagons are more accurate, and thus more time is spend during the consistency but we need to split less.

As for the last table containing results for this comparison, Table 4.3, it compares the number of solution boxes to the number of solution octagons. We can see that for two problems `pramanik` and `trigo1`, the number of octagons solution is larger than the number of solution boxes. For these two problems, octagons have trouble with the consistency for the elements on the limit of the solution set.

| Name | # var | ctr type | $\mathbb{B}$ | $\mathbb{O}$ | CB | R | SL | P |
|---|---|---|---|---|---|---|---|---|
| brent-10 | 10 | = | **21.58** | 330.73 | 89.78 | 92.59 | 105.91 | 109.74 |
| o32 | 5 | $\leq$ | 27.25 | 40.74 | 40.74 | **17.63** | 20.68 | 21.23 |
| ipp | 8 | = | 38.83 | 279.36 | 279.36 | 30.14 | **29.07** | 41.60 |
| trigo1 | 10 | = | 40.23 | 253.53 | 253.53 | 38.60 | **37.03** | **37.03** |
| KinematicPair | 2 | $\leq$ | **59.04** | 60.78 | 60.78 | 60.78 | 60.78 | 60.78 |
| nbody5.1 | 6 | = | 95.99 | 27.08 | **22.13** | 443.07 | 50.87 | 439.69 |
| bellido | 9 | = | **111.12** | - | - | 362.69 | 361.56 | 318.09 |
| pramanik | 3 | = | 281.80 | 240.96 | 240.96 | 141.94 | 137.42 | **131.30** |
| caprasse | 4 | = | 9175.36 | - | - | **1085.21** | 1131.33 | 2353.58 |
| h74 | 4 | = $\leq$ | - | 647.76 | 647.76 | - | **0.15** | **0.15** |

**Table 4.5.** *Results on problems from the Coconut benchmark. For each problem, we give the CPU time in seconds to solve it using the intervals ($\mathbb{B}$), the octagons ($\mathbb{O}$) and all the partial octagons, Constraint Based (CB), Random (R), Strongest-Link (SL) and Promising (P)*

| Name | # var | ctr type | $\mathbb{B}$ | $\mathbb{O}$ | CB | R | SL | P |
|---|---|---|---|---|---|---|---|---|
| brent-10 | 10 | = | 211 885 | **5 467** | 5 841 | 175 771 | 205 485 | 21 495 |
| o32 | 5 | $\leq$ | 161 549 | **25 319** | **25 319** | 52 071 | 62 911 | 72 565 |
| ipp | 8 | = | 237 445 | **21 963** | **21 963** | 51 379 | 50 417 | 75 285 |
| trigo1 | 10 | = | 13 621 | **4 425** | **4 425** | 6 393 | 5 943 | 5 943 |
| KinematicPair | 2 | $\leq$ | 847 643 | **373 449** | **373 449** | **373 449** | **373 449** | **373 449** |
| nbody5.1 | 6 | = | 598 521 | 5 435 | **5 429** | 578 289 | 137 047 | 542 263 |
| bellido | 9 | = | 774 333 | - | - | 577 367 | 573 481 | **496 729** |
| pramanik | 3 | = | 1 992 743 | **243 951** | **243 951** | 346 633 | 315 861 | 319 037 |
| caprasse | 4 | = | 150 519 891 | - | - | **4 445 655** | 4 472 839 | 9 933 597 |
| h74 | 4 | = $\leq$ | - | 418 867 | 418 867 | - | **625** | **625** |

**Table 4.6.** *Results on problems from the Coconut benchmark. For each problem, we give the number of box created while solving with the intervals ($\mathbb{B}$), as well as the number of octagons created while solving with the octagons ($\mathbb{O}$) and all the partial octagons, Constraint Based (CB), Random (R), Strongest-Link (SL) and Promising (P)*

#### 4.4.4.2. *Comparison of heuristics for the choice of variable*

Table 4.4 compares the results, in terms of CPU time, of the different variable heuristics. We can see that splitting the largest size domain (LF) is not a good strategy. Indeed, if the domains are not well-propagated, this may signify that the basis in which these domains are, is of little interest for the problem we try to solve. Restricting the choice to variables living in rotated basis (LOF) does not improve the results and even deteriorates them. This shows that there exist bases in which the domains are large and do not add relevant information. Moreover, this means that sometimes it is necessary to split in the canonical basis.

| Name | # var | ctr type | $\mathbb{B}$ | $\mathbb{O}$ | CB | R | SL | P |
|---|---|---|---|---|---|---|---|---|
| brent-10 | 10 | $=$ | 825 | **149** | 153 | 636 | 1 643 | 1 765 |
| o32 | 5 | $\leq$ | 74 264 | **12 523** | **12 523** | 35 868 | 31 216 | 25 833 |
| ipp | 8 | $=$ | 2 301 | **2243** | **2243** | 4 329 | 6 922 | 6 194 |
| trigo1 | 10 | $=$ | 40 | **32** | **32** | 168 | 140 | 140 |
| KinematicPair | 2 | $\leq$ | 346 590 | **186 717** | **186 717** | **186 717** | **186 717** | **186 717** |
| nbody5.1 | 6 | $=$ | 1 012 | 1 003 | **996** | 1 794 | 50 526 | 2 230 |
| bellido | 9 | $=$ | **7 372** | - | - | 34 756 | 32 482 | 37 508 |
| pramanik | 3 | $=$ | 149 011 | **54 659** | **54 659** | 72 052 | 69 621 | 71 239 |
| caprasse | 4 | $=$ | 1 544 | - | - | 164 | 148 | **110** |
| h74 | 4 | $=$ $\leq$ | - | 209 406 | 209 406 | - | **293** | **293** |

**Table 4.7.** *Results on problems from the Coconut benchmark. For each problem, we give the number of box in the computed solution by the intervals ($\mathbb{B}$), as well as the number of octagons in the solution computed with the octagons ($\mathbb{O}$) and all the partial octagons, Constraint-Based (CB), Random (R), Strongest-Link (SL) and Promising (P)*

When the choice of variables is restricted to the variables in the canonical basis (LCF), the results are much better and the time limit is not exceeded for any of the selected problems. This can be explained by the fact that all the rotated bases depend on the canonical basis. And therefore, any change in the canonical basis is passed on several other basis and may improve the approximation.

Finally, the strategy Oct-Split (OS) is more often the best. By splitting in a basis of interest, this heuristic efficiently explores the search space. In addition, it suits more to the definition of the precision function, which is part of the termination criterion. For these reasons, we used this heuristic for the octagonal solving in the following experiments.

#### 4.4.4.3. *Comparison of octagonalization heuristics*

By looking at the computation time in Table 4.5, octagons and partial octagons obtain very poor results on problems `brent-10` and `bellido`. In these two problems, the rotated bases have very few rotated constraints. This low number of rotated constraints does not help to reduce the domains of the rotated variables, and thus, does not help in reducing the domains of the initial variables. As, the rotated variables are linked to the initial variables, reducing the first, reduces the latter and conversely. Conversely, if the rotated variables are not reduced, no information will be given to the initial variables. In other words, the information gain provided by these rotated constraints and variables do not compensate for the time spent to propagate them.

However for problems `h74`, `caprasse` and `nbody5.1`, the octagonal resolution or one of the resolutions using partial octagons, highly improves the results. Given the form of the constraints, we were hoping to improve the computation time on these problems. Problem `h74` contains constraints of the form $v_i - v_j \leq c$, for only one pair $(i, j)$, which correspond to octagonal constraints. Thus, for the $(i, j)$-rotated basis, these constraints are well-propagated and solutions are quickly found. Note that, for this problem, when all the bases are generated, the computation time is large. This shows that the resolution spend a lot of time trying to reduce domains of variables living in bases of little interest, which do not add information. The situation is the same for problem `caprasse`. On the contrary, in problem `nbody5.1`, all the variables can be grouped in pairs following the promising pattern (definition 4.9). Therefore, generating all the bases ($\mathbb{O}$) or just a subset corresponding to variables pairs (CB) highly improves the results. Moreover, we can see that one of the pair is more important than the others, because the results obtained with the Strongest-Link (SL) heuristic are not deteriorated. We can also say that this pair does not correspond to the pair with the biggest number of promising patterns, as the Promising (P) heuristic obtain very poor results. Finally, it seems that it corresponds to the basis maximizing the number of rotated constraints.

For problem `KinematicPair`, the time obtained with the octagons is equivalent to the time obtained with the intervals. For the other problems, the time obtained by one of the solving methods using partial octagons slightly improve the computation time obtained with the intervals.

Still looking at the computation time, but this time only for the results obtained for each of the octagonalization heuristics as a whole rather that on a case to case basis for each problem, we can see that, contrary to what was expected, it is the Strongest-Link (SL) octagonalization heuristic which gives the best results. We expected to have better results with the Promising (P) heuristic. We can therefore deduce that the more rotated constraints is, the better the propagation of the rotated variables is. This support the hypothesis given to explain the poor results on problems `brent-10` and `bellido`.

In Table 4.6, we can see that the number of octagons created during the resolution process is always smaller than the number of boxes created. Moreover, in most of the problems, the more bases are generated for the octagons, the less octagons are created during the resolution. It follows that, the more precise the representation is, the more time is spent on the propagation and fewer splits are needed.

The conclusion is the same for Table 4.6. The number of octagons solution is usually smaller when the octagons are composed of a large number of bases.

## 4.5. Conclusion

In this chapter, we have implemented the unified solving method introduced in Chapter 2 with the octagon abstract domain (Chapter 3). We gave possible algorithms for the different components of the octagonal solving process, such as the consistency and propagation scheme. Moreover, we proposed different heuristic for the choice of the domain to split and the creation of partial octagons. The details of a solver based on Ibex are given. And preliminary results on a classical benchmark for continuous problems are given. These results are encouraging and show that an octagonal splitting operator and exploration strategy are mandatory to better take into account the relations within an octagon and obtain a better octagonal solving method. Several perspectives are possible such as the use of recent propagators less sensitive of multiple occurrences of variables like Mohc [ARA 10]; the definition of octagonalization heuristic offering a better compromise between SL and Constraint-Based (CB); the definition of other abstract domains in CP such as the polyhedra [COU 78] or interval polyhedra [CHE 09].

# 5

# An Abstract Solver: AbSolute

In the previous chapters, we have defined some notions of abstract interpretation (AI) in constraint programming (CP), allowing us to define a unified resolution scheme that does not depend on the chosen abstract domain. This scheme has been implemented giving us an octagonal solver. This early work makes it possible, in theory, to develop a mixed solver. However, there is a practical obstacle: the lack of representation for both integer and real variables in solvers in CP. A solution is to do the opposite: use abstract interpretation abstract domains that do not depend on the variables type, and add the CP resolution process in AI. In this chapter, we present CP using only notions from AI. We then present an abstract solver, AbSolute, implemented on top of an abstract domain library.

Using the links highlighted in section 1.3.1, we express the constraints resolution in AI. Solving a constraint satisfaction problem (CSP) is seen as a process of concrete semantics. We thus define concrete and abstract domains, abstract operators for splits and consistency and an iterative scheme computing an approximation of its concrete semantics. These definitions allow us to obtain an abstract solving method. We have implemented and tested this method on different problems. As for the definitions named above, the details of this implementation and an analysis of the results obtained are given in this chapter.

## 5.1. Abstract solving method

Similarly to the analysis of a program in AI, we define the concrete semantics of solving a constraint satisfaction problem. Then, we show that the representations used in CP correspond to abstract domains with a size function. In addition, we show that the lower closure operators in AI are similar to the propagators in CP, and therefore local iterations are equivalent to the propagation loop. Splitting operators do not exist in AI, we define them and thus obtain an abstract solving method.

### 5.1.1. *Concrete solving as concrete semantics*

Previously, in section 1.2.1, we saw that solving a constraint satisfaction problem is similar to computing the solutions set of a conjunction of constraints. Given values for the variables, constraints answer true or false in the case of discrete variables, and true, false or maybe for continuous variables. From this observation, solving a constraint satisfaction problem can be seen as the analysis of a conjunction of tests, the constraints. The concrete semantic thus corresponds to the solutions set of the problem. We consider as concrete domain $\mathcal{D}^\flat$ the subsets of the initial search space $\hat{D} = \hat{D}_1 \times \cdots \times \hat{D}_n$ (definition 1.9), that is $(\mathcal{P}(\hat{D}), \subseteq, \emptyset, \cup)$. Similarly, in CP, each constraint $C_i$ is associated with a propagator, which corresponds in AI to a lower closure operator (definition 1.8) $\rho_i^\flat : \mathcal{P}(\hat{D}) \rightarrow \mathcal{P}(\hat{D})$, such that $\rho_i^\flat(X)$ only keeps points of $X$ satisfying constraint $C_i$. Finally, a concrete solution to the problem is simply $S = \rho^\flat(\hat{D})$, where $\rho^\flat = \rho_1^\flat \circ \cdots \circ \rho_p^\flat$.

Furthermore, by formalizing the resolution of a constraints satisfaction problem in terms of local iterations, the solutions can be expressed as a fixpoint $\mathrm{gfp}_{\hat{D}}\, \rho^\flat$, which is the greatest fixpoint of the composition of propagators, $\rho^\flat$, smaller than the initial search space, $\hat{D}$. Solving a problem is equivalent to computing a fixpoint. Expressing the solution set as the greatest fixpoint has already been made [SCH 05].

The solutions set being identified as the concrete domain, existing representations for domains in CP can be seen as abstract domains.

### 5.1.2. *Abstract domains existing in CP*

Solvers do not manipulate individual points in $\hat{D}$, but rather collections of points of certain forms, such as boxes, called domains in CP. We now show that CP domains can be the base set $\mathcal{D}^\sharp$ of an abstract domain $(\mathcal{D}^\sharp, \sqsubseteq^\sharp, \perp^\sharp, \sqcup^\sharp)$ in AI.

DEFINITION 5.1 (Integer Cartesian product).– *Let* $v_1, \ldots, v_n$ *be variables over finite discrete domains* $\hat{D}_1 \ldots \hat{D}_n$. *We call integer Cartesian product any Cartesian product of integer sets and is expressed in* $\mathcal{D}^\flat$ *by :*

$$\mathcal{S}^\sharp = \left\{ \prod_i X_i \mid \forall i, X_i \subseteq \hat{D}_i \right\}$$

*This definition corresponds to definition 1.10.*

DEFINITION 5.2 (Integer box).– *Let* $v_1, \ldots, v_n$ *be variables over finite discrete domains* $\hat{D}_1 \ldots \hat{D}_n$. *We call integer box a Cartesian product of integer intervals and is expressed in* $\mathcal{D}^\flat$ *by :*

$$\mathcal{I}^\sharp = \left\{ \prod_i [\![a_i, b_i]\!] \mid \forall i,\ [\![a_i, b_i]\!] \subseteq \hat{D}_i,\ a_i \leq b_i \right\} \cup \emptyset$$

*This definition corresponds to definition 1.11.*

DEFINITION 5.3 (Box).– *Let* $v_1, \ldots, v_n$ *be variables over bounded continuous domains* $\hat{D}_1 \ldots \hat{D}_n$. *A box is a Cartesian product of intervals and is expressed in* $\mathcal{D}^\flat$ *by :*

$$\mathcal{B}^\sharp = \left\{ \prod_i I_i \mid I_i \in \mathbb{I}, I_i \subseteq \hat{D}_i \right\} \cup \emptyset$$

*This definition corresponds to definition 1.12.*

### 5.1.3. *Abstract domains operators*

In CP, to each representation is associated a consistency. Along the same lines, we associate with each abstract domain a consistency, and in addition to the standard operators in AI, we define a monotonic *size function* $\tau : \mathcal{D}^\sharp \to \mathbb{R}^+$, which is used as a termination criterion for the splitting operator (definition 5.7).

EXAMPLE 5.1 (Integer Cartesian products abstract domain $\mathcal{S}^\sharp$).– Generalized arc-consistency (definition 1.15) corresponds to the abstract domain of integer Cartesian products $\mathcal{S}^\sharp$ (definition 5.1), ordered by element-wise set inclusion. It is linked with the concrete domain $\mathcal{D}^\flat$ by the standard Cartesian Galois connection:

$$\mathcal{D}^\flat \underset{\alpha_a}{\overset{\gamma_a}{\leftrightarrows}} \mathcal{S}^\sharp$$

$$\gamma_a(S_1, \ldots, S_n) = S_1 \times \cdots \times S_n$$
$$\alpha_a(X^\flat) = \lambda i. \{v \mid \exists (x_1, \ldots, x_n) \in X^\flat,\, x_i = v\}$$

The size function $\tau_a$ uses the size of the largest component:

$$\tau_a(S_1, \ldots, S_n) = \max_i(|S_i|)$$

Thus, if the considered element is a solution, all variables are instantiated, so all sets are singletons and $\tau_a$ is equal to 1.

EXAMPLE 5.2 (Integer boxes abstract domain $\mathcal{I}^\sharp$).– Bound consistency (definition 1.16) corresponds to the abstract domain of integer boxes $\mathcal{I}^\sharp$ (definition 5.2), ordered by element-wise interval inclusion. We have a Galois connection:

$$\mathcal{D}^\flat \underset{\alpha_b}{\overset{\gamma_b}{\leftrightarrows}} \mathcal{I}^\sharp$$

$$\gamma_b([\![a_1, b_1]\!], \ldots, [\![a_n, b_n]\!]) = [\![a_1, b_1]\!] \times \cdots \times [\![a_n, b_n]\!]$$
$$\alpha_b(X^\flat) = \lambda i. [\![ \min \{v \in \mathbb{Z} \mid \exists (x_1, \ldots, x_n) \in X^\flat,\, x_i = v\},$$
$$\max \{v \in \mathbb{Z} \mid \exists (x_1, \ldots, x_n) \in X^\flat,\, x_i = v\} ]\!]$$

We use the length of the largest dimension plus one as a size function so that, like the integer Cartesian products, if the element is a solution, the size function is equal to 1:

$$\tau_b([\![a_1, b_1]\!], \ldots, [\![a_n, b_n]\!]) = \max_i(b_i - a_i) + 1$$

EXAMPLE 5.3 (Boxes abstract domain $\mathcal{B}^\sharp$).– Hull consistency (definition 1.17) corresponds to the abstract domain of boxes with floating-point bounds $\mathcal{B}^\sharp$ (definition 5.3). We use the following Galois connection:

$$\mathcal{D}^\flat \underset{\alpha_h}{\overset{\gamma_h}{\leftrightarrows}} \mathcal{B}^\sharp$$

$$\gamma_h([a_1, b_1], \ldots, [a_n, b_n]) = [a_1, b_1] \times \cdots \times [a_n, b_n]$$
$$\alpha_h(X^\flat) = \lambda i.[\max\{v \in \mathbb{F} \mid \forall(x_1, \ldots, x_n) \in X^\flat,\ x_i \geq v\}, \min\{v \in \mathbb{F} \mid \forall(x_1, \ldots, x_n) \in X^\flat,\ x_i \leq v\}]$$

The size function corresponds to the size of the largest dimension:

$$\tau_h([a_1, b_1], \ldots, [a_n, b_n]) = \max_i(b_i - a_i)$$

We observe that to each choice corresponds a classic non-relational abstract domain in AI. Moreover, it is an homogeneous Cartesian product of identical single-variable domains (representation for one variable). However, this does not need to be the case, and new solvers can be designed beyond the ones considered in traditional CP by varying the abstract domains further. A first idea is to apply different consistencies to different variables which permits, in particular, mixing variables with discrete domains and variables with continuous domains. Yet, the domains still corresponds to Cartesian products. A second idea is to parameterize the solver with other abstract domains from the AI literature, in particular relational or mixed domains. We choose the second idea, so the representations are no longer restricted to Cartesian products and can represent different types of variables. This idea is illustrated below.

EXAMPLE 5.4 (Octagon abstract domain $\mathcal{O}^\sharp$).– The octagon abstract domain $\mathcal{O}^\sharp$ [MIN 06] assigns a (floating-point) upper bound to each binary unit expression $\pm v_i \pm v_j$ on the variables $v_1, \ldots, v_n$.

$$\mathcal{O}^\sharp = \{\alpha v_i + \beta v_j \mid i, j \in [\![1, n]\!],\ \alpha, \beta \in \{-1, 1\}\} \to \mathbb{F}$$

It enjoys a Galois connection with $\mathcal{D}^\flat$. Let us recall here this Galois connection and the size function defined previously (definition 3.7). $\forall X^\sharp \in \mathcal{O}^\sharp, \forall i, j, \in [\![1, n]\!], \forall \alpha, \beta \in \{-1, 1\}$, we denote by $X^\sharp(\alpha v_i + \beta v_j)$ the floating-point upper bound of the binary unit expression $\alpha v_i + \beta v_j$.

$$\mathcal{D}^\flat \underset{\alpha_o}{\overset{\gamma_o}{\leftrightarrows}} \mathcal{O}^\sharp$$

$$\gamma_o(X^\sharp) = \{(x_1, \ldots, x_n) \mid \forall i, j, \alpha, \beta,\ \alpha x_i + \beta x_j \leq X^\sharp(\alpha v_i + \beta v_j)\}$$

$$\alpha_o(X^\flat) = \lambda(\alpha v_i + \beta v_j).\max\{\alpha x_i + \beta x_j \mid (x_1, \ldots, x_n) \in X^\flat\}$$

$$\tau_o(X^\sharp) = \min\Big(\max_{i,j,\beta}\,(X^\sharp(v_i + \beta v_j) + X^\sharp(-v_i - \beta v_j)),\ \max_{i}\,(X^\sharp(v_i + v_i) + X^\sharp(-v_i - v_i))/2\Big)$$

EXAMPLE 5.5 (Polyhedron abstract domain $\mathcal{P}^\sharp$).– The polyhedron domain $\mathcal{P}^\sharp$ [COU 78] abstracts sets as convex, closed polyhedra. Modern implementations [JEA 09] generally follow the "double description approach" and maintain two dual representations for each polyhedron: a set of linear constraints and a set of generators. A generator is either a vertex or a ray of the polyhedron. A ray corresponds to a vector along which, starting from any vertex of the polyhedron, any point is part of the polyhedron. However, the polyhedra we used do not have rays, given that they are bounded. There is no abstraction function $\alpha$ for polyhedra, and therefore no Galois connection. Operators are generally easier in one representation. In particular, we define the size function on generators as the maximal Euclidian distance between pairs of vertices. Let $X^\sharp \in \mathcal{P}^\sharp$ :

$$\tau_p(X^\sharp) = \max_{g_i, g_j \in X^\sharp} ||g_i - g_j||$$

EXAMPLE 5.6 (Mixed boxes abstract domain $\mathcal{M}^\sharp$).– The mixed box abstract domain $\mathcal{M}^\sharp$, ordered by inclusion. Let $v_1 \ldots v_n$ be the variables set and let $(v_1, \ldots, v_m), m \in [\![1, n]\!]$ the set of integer variables and $(v_{m+1}, \ldots, v_n)$ the set of real variables. The abstract domain of mixed boxes assigns to integer variables an integer interval and to real variables a real interval with floating-point bounds.

$$\mathcal{M}^\sharp = \left\{ \prod_{i=1}^{m} [\![a_i, b_i]\!] \mid \forall i,\ [\![a_i, b_i]\!] \subseteq \hat{D}_i,\ a_i \leq b_i \right\}$$
$$\times \left\{ \prod_{i=m+1}^{n} I_i \mid I_i \in \mathbb{I}, I_i \subseteq \hat{D}_i \right\} \cup \emptyset$$

This abstract domain enjoys a Galois connection with $\mathcal{D}^\flat$:

$$\mathcal{D}^\flat \underset{\alpha_m}{\overset{\gamma_m}{\leftrightarrows}} \mathcal{M}^\sharp$$

$$\gamma_m([\![a_1, b_1]\!], \ldots, [\![a_m, b_m]\!], [a_{m+1}, b_{m+1}], \ldots, [a_n, b_n])$$
$$= [\![a_1, b_1]\!] \times \cdots \times [\![a_m, b_m]\!] \times [a_{m+1}, b_{m+1}] \times \cdots \times [a_n, b_n]$$

$$\alpha_m(X^\flat) = \lambda i. \begin{cases} [\![\ \min \{v \in \mathbb{Z} \mid \exists (x_1, \ldots, x_n) \in X^\flat,\ x_i = v\}, & i \in [\![1, m]\!] \\ \quad \max \{v \in \mathbb{Z} \mid \exists (x_1, \ldots, x_n) \in X^\flat,\ x_i = v\}\ ]\!] & \\ [\ \max \{v \in \mathbb{F} \mid \forall (x_1, \ldots, x_n) \in X^\flat,\ x_i \geq v\}, & i \in [\![m+1, n]\!] \end{cases}$$

The size function corresponds to the size of the largest dimension:

$$\tau_m([\![a_1, b_1]\!], \ldots, [\![a_m, b_m]\!], [a_{m+1}, b_{m+1}], \ldots, [a_n, b_n]) = \max_i (b_i - a_i)$$

### 5.1.4. *Constraints and consistency*

We now assume that an abstract domain $\mathcal{D}^\sharp$ underlying the solver is fixed.

Given the concrete semantics of the constraints $\rho^\flat = \rho_1^\flat \circ \cdots \circ \rho_p^\flat$, and if $\mathcal{D}^\sharp$ enjoys a Galois connection $\mathcal{D}^\flat \underset{\alpha}{\overset{\gamma}{\leftrightarrows}} \mathcal{D}^\sharp$, then the semantics of the perfect propagator achieving the consistency for all the constraints is simply: $\alpha \circ \rho^\flat \circ \gamma$. Solvers achieve this algorithmically. They apply the propagator for each constraint, in turn, until a fixpoint is reached or, when this process is deemed too costly, return before a fixpoint is reached.

REMARK 5.1.– By observing that each propagator corresponds to an abstract test transfer function $\rho_i^\sharp$ in $\mathcal{D}^\sharp$, we retrieve the local iterations proposed by Granger to analyze conjunctions of tests [GRA 92]. A trivial narrowing is used here: stop refining after an iteration limit is reached.

Additionally, each $\rho_i^\sharp$ can be internally implemented by local iterations [GRA 92], a technique which is used in both the AI and CP communities. A striking connection is the analysis in non-relation domains using forward–backward iterations on expression trees [MIN 04, section 2.4.4], which is extremely similar to the HC4-Revise algorithm [BEN 99] developed independently for CP.

We can thus see the consistency as an abstraction of the concrete semantics of constraints.

### 5.1.5. *Disjunctive completion and split*

To approximate the solutions to an arbitrary precision, solvers use a coverage of finite abstract elements from $\mathcal{D}^\sharp$. This corresponds in AI to the notion of disjunctive completion [COU 92a].

DEFINITION 5.4 (Disjunctive completion).– *Let $\mathcal{D}^\sharp$ be a set. A disjunctive completion $\mathcal{E}^\sharp = \mathcal{P}_{\text{finite}}(\mathcal{D}^\sharp)$ is a subset of $\mathcal{D}^\sharp$ whose elements are not comparable. That is:*

$$\mathcal{E}^\sharp = \{X^\sharp \subseteq \mathcal{D}^\sharp \mid \forall B^\sharp, C^\sharp \in X^\sharp, B^\sharp \subsetneq^\sharp C^\sharp\}$$

EXAMPLE 5.7 (Disjunctive completion).– All the elements at the same level in a lattice form a disjunctive completion as they cannot be compare to each other. Consider the lattice of integer boxes $\mathcal{I}^\sharp$ with inclusion, the set

$$\{[\![1,2]\!] \times [\![1,2]\!], [\![1,2]\!] \times [\![3,7]\!]\}$$

is a disjunctive completion. Indeed, the integer box $[\![1,2]\!] \times [\![1,2]\!]$ is not comparable with the inclusion to the integer box $[\![1,2]\!] \times [\![3,7]\!]$. The first box is not included in the second and vice versa.

We now consider the abstract domain $\mathcal{E}^\sharp = \mathcal{P}_{\text{finite}}(\mathcal{D}^\sharp)$, and equip it with the Smyth order $\sqsubseteq^\sharp_\mathcal{E}$, such that $\forall X^\sharp, Y^\sharp \in \mathcal{E}^\sharp$ two disjunctive completions, $X^\sharp$ is included in $Y^\sharp$ if and only if each element of $X^\sharp$ is included in one of the elements of $Y^\sharp$. It is a classic order for disjunctive completions defined as:

$$X^\sharp \sqsubseteq^\sharp_\mathcal{E} Y^\sharp \iff \forall B^\sharp \in X^\sharp, \exists C^\sharp \in Y^\sharp, B^\sharp \sqsubseteq^\sharp C^\sharp$$

EXAMPLE 5.8 (Smyth order).– Consider the lattice of boxes $\mathcal{B}^\sharp$ with inclusion. Let $X^\sharp = \{[2,5] \times [1,4.5], [1,4] \times [4.5,8], [5,8] \times [2,5.5], [6,9] \times [5.5,9]\}$ and $Y^\sharp = \{[1,5] \times [1,8], [5,9] \times [2,9]\}$ two disjunctive completions. We have $X^\sharp \sqsubseteq^\sharp_\mathcal{E} Y^\sharp$. Indeed, each box of $X^\sharp$ is included in a box of $Y^\sharp$. Graphically, we have the following figure; the boxes on the left are included in the boxes on the right:

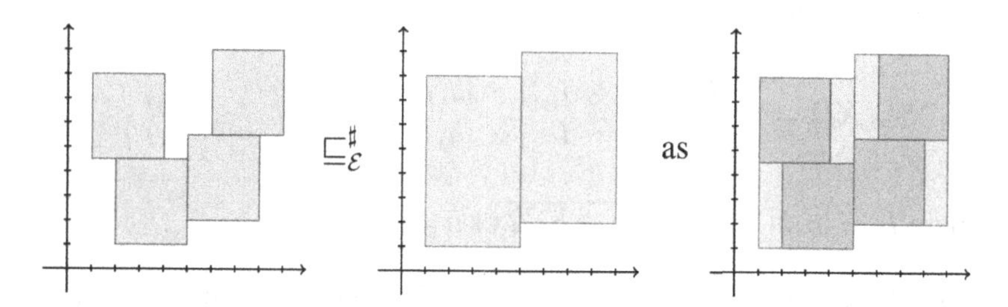


The splitting operator $\oplus$ splits an element of $\mathcal{D}^\sharp$ in two or more elements. It thus transforms an element of $\mathcal{D}^\sharp$ into an element of $\mathcal{E}^\sharp$. So, it achieves the creation of new disjunctive completions. Since the splitting operator does not exist in AI, we redefine it here.

DEFINITION 5.5 (Split operator).– *A* split operator *is an operator* $\oplus : \mathcal{D}^\sharp \to \mathcal{E}^\sharp$ *such that* $\forall e \in \mathcal{D}^\sharp$*:*

1) $| \oplus (e)|$ *is finite;*

2) $\forall e_i \in \oplus(e),\ e_i \sqsubseteq^\sharp e$*;*

3) $\gamma(e) = \bigcup\{\gamma(e_i) \mid e_i \in \oplus(e)\}$.

*This definition is the abstract domains version of definition 2.3.*

Each element of $\oplus(e)$ is included in $e$ (condition 5.5), and we have $\oplus(e) \sqsubseteq^\sharp_\mathcal{E} \{e\}$. Furthermore, condition 5.5 shows that $\oplus$ is an abstraction of the identity. Thus, $\oplus$ can be freely applied at any place during the solving process without altering the soundness. The integer instantiation and the split on boxes can be retrieved with this definition.

EXAMPLE 5.9 (Split in $\mathcal{S}^\sharp$).– The instantiation of a variable $v_i$ in a discrete domain $X^\sharp = (S_1, \ldots, S_n) \in \mathcal{S}^\sharp$ is a splitting operator:

$$\oplus_a(X^\sharp) = \{(S_1, \ldots, S_{i-1}, x, S_{i+1}, \ldots, S_n) \mid x \in S_i\}$$

EXAMPLE 5.10 (Cut into $\mathcal{I}^\sharp$).– The instantiation of a variable $v_i$ in a discrete domain $X^\sharp = (X_1, \ldots, X_n) \in \mathcal{I}^\sharp$ is a splitting operator:

$$\oplus_b(X^\sharp) = \{(X_1 \times \cdots \times X_{i-1} \times x \times X_{i+1} \times \cdots \times X_n) \mid x \in X_i\}$$

EXAMPLE 5.11 (Split in $\mathcal{B}^\sharp$).– Cutting a box into a variable $v_i$ in a continuous domain $X^\sharp = (I_1, \ldots, I_n) \in \mathcal{B}^\sharp$ is a splitting operator:

$$\oplus_h(X^\sharp) = \begin{array}{l} \{\ (I_1 \times \cdots \times I_{i-1} \times [a, h] \times I_{i+1} \times \cdots \times I_n), \\ \quad (I_1 \times \cdots \times I_{i-1} \times [h, b] \times I_{i+1} \times \cdots \times I_n)\ \} \end{array}$$

where $I_i = [a, b]$ and $h = \overline{(a + b)/2}$ or $h = \underline{(a + b)/2}$.

Other splitting operators can be defined for non-relational or mixed abstract domains.

EXAMPLE 5.12 (Split in $\mathcal{O}^\sharp$).– Given a binary unit expression $\alpha v_i + \beta v_j, i, j \in [\![1, n]\!], \alpha, \beta \in \{-1, 1\}$, we define the split on an octagon $X^\sharp \in \mathcal{O}^\sharp$ along this expression as:

$$\oplus_o(X^\sharp) = \{X^\sharp[(\alpha v_i + \beta v_j) \mapsto h],\ X^\sharp[(-\alpha v_i - \beta v_j) \mapsto -h]\}$$

where $h = (X^\sharp(\alpha v_i + \beta v_j) - X^\sharp(-\alpha v_i - \beta v_j))/2$, rounded in $\mathbb{F}$ in any direction.

This operator corresponds to the one presented in definition 3.6. Indeed, cutting along a binary unit expression $\alpha v_i + \beta v_j$ is equivalent to splitting the domain of variable $v_i^{i,j}$ or $v_j^{i,j}$ when $i \neq j$, and of variable $v_i$ or $v_j$ when $i = j$.

EXAMPLE 5.13 (Split in $\mathcal{P}^\sharp$).– Given a polyhedron $X^\sharp \in \mathcal{P}^\sharp$ represented as a set of linear constraints, and a linear expression $\sum_i \beta_i v_i$, we define the splitting operator on a polyhedron along this expression as:

$$\oplus_p(X^\sharp) = \left\{ X^\sharp \cup \left\{ \sum_i \beta_i v_i \leq h \right\}, X^\sharp \cup \left\{ \sum_i \beta_i v_i \geq h \right\} \right\}$$

where $h = \frac{\left( \min\limits_{\gamma(X^\sharp)} \sum\limits_i \beta_i v_i + \max\limits_{\gamma(X^\sharp)} \sum\limits_i \beta_i v_i \right)}{2}$ can be computed by the Simplex algorithm [DAN 97].

EXAMPLE 5.14 (Split in $\mathcal{M}^\sharp$).– Consider the mixed box $X^\sharp = (X_1, \ldots, X_m, I_{m+1}, \ldots, I_n) \in \mathcal{M}^\sharp$. We define the splitting operator on a mixed box, such that either a discrete variable is instantiated, or the domain of a continuous variable is split:

$$\oplus_m(X^\sharp) = \begin{cases} \{(X_1 \times \cdots \times X_{i-1} \times x \times X_{i+1} \times \cdots \times X_m \times I_{m+1} \times \cdots \times I_n) \\ \quad | \, x \in X_i\} & \text{if } i \in [\![1, m]\!] \\ \{(X_1 \times \ldots X_m \times I_{m+1} \times \cdots \times I_{i-1} \times [a, h] \times I_{i+1} \times \cdots \times I_n), \\ \quad (X_1 \times \ldots X_m \times I_{m+1} \times \cdots \times I_{i-1} \times [h, b] \times I_{i+1} \times \cdots \times I_n) \,\} & \text{if } i \in [\![m+1, n]\!] \end{cases}$$

where $\forall i \in [\![m+1, n]\!], I_i = [a, b]$ and $h = \overline{(a+b)/2}$ or $h = \underline{(a+b)/2}$.

These splitting operators are parameterized by the choice of a direction of cut (some variable or expression). For non-relational domains, we can use two classic strategies from CP: define and order and split each variable, in turn, or split along a variable with maximal size or with the smallest domain; that is $|S_i|$ for a variable represented with a set or $b - a$ for a variable represented with an interval $[a, b]$ or $[\![a, b]\!]$. These strategies lift naturally to octagons by replacing the set of variables with the (finite) set of unit binary expressions. For polyhedra, we can bisect the segment between two vertices that are the farthest apart, in order to minimize $\tau_p$. However, even for relational domains, we can use a faster and simpler non-relational split, such as cutting along the variable with the largest domain.

DEFINITION 5.6 (Choice operator).– *A* choice operator *is an operator* $\pi : \mathcal{E}^\sharp \to \mathcal{D}^\sharp$ *such that* $\forall X^\sharp \in \mathcal{E}^\sharp$, *for* $r \in \mathbb{R}^{>0}$ *fixed:*

1) $\pi(X^\sharp) \in X^\sharp$ *and*

2) $(\tau \circ \pi)(X^\sharp) > r$.

*The choice operator chose an element in the disjunctive completion* $X^\sharp$, *bigger than a given value* $r$.

To ensure the termination of the solver, we impose that any series of reductions, splits and choices eventually outputs a small enough element for $\tau$.

DEFINITION 5.7 (Compatibility of $\tau$ and $\oplus$).– *The two operators* $\tau$: $\mathcal{D}^\sharp \to \mathcal{R}^+$ *and* $\oplus : \mathcal{D}^\sharp \to \mathcal{E}^\sharp$ *are said compatible if and only if, for any reductive operator* $\rho^\sharp : \mathcal{D}^\sharp \to \mathcal{D}^\sharp$ *(i.e.* $\forall X^\sharp \in \mathcal{D}^\sharp, \rho^\sharp(X^\sharp) \sqsubseteq^\sharp X^\sharp$*) and any family of choice operators* $\pi_i : \mathcal{E}^\sharp \to \mathcal{D}^\sharp$ *(i.e.* $\forall Y^\sharp \in \mathcal{E}^\sharp, \pi_i(Y^\sharp) \in Y^\sharp$*), we have:*

$$\forall e \in \mathcal{D}^\sharp, \forall r \in \mathbb{R}^{>0}, \exists K \text{ such that}$$

$$\forall j \geq K, (\tau \circ \pi_j \circ \oplus \circ \rho^\sharp \circ \cdots \circ \pi_1 \circ \oplus \circ \rho^\sharp)(e) \leq r$$

All operators $\oplus$ and $\rho^\sharp$ being contracting, the operator $\pi \circ \oplus \circ \rho^\sharp$ is also reducing the domains, and thus for any $r$ strictly positive, the previous definition is verified.

We can easily verify that the splitting operators previously introduced, $\oplus_a$, $\oplus_b$, $\oplus_h$, $\oplus_o$, $\oplus_p$ and $\oplus_m$, are, respectively, compatible with the size functions, $\tau_a$, $\tau_b$, $\tau_h$, $\tau_o$, $\tau_p$ and $\tau_m$, respectively, proposed for the $\mathcal{S}^\sharp$, $\mathcal{I}^\sharp$, $\mathcal{B}^\sharp$, $\mathcal{O}^\sharp$, $\mathcal{P}^\sharp$ and $\mathcal{M}^\sharp$ abstract domain.

REMARK 5.2.– The search procedure can be represented as a search tree (as defined in section 1.2.4). With this representation, the set of nodes at a given depth corresponds to a disjunction overapproximating the solutions set. Moreover, a series of reduction ($\rho$), selection ($\pi$) and split ($\oplus$) operators corresponds to a tree branch. Definition 5.7 states that each branch of the search tree is finite.

### 5.1.6. *Abstract solving*

The abstract solving algorithm is given in algorithm 5.1. It maintains in toExplore and sols two disjunctions in $\mathcal{E}^\sharp$, and iterates the following steps:

1) choose an abstract element $e$ in toExplore (*pop*);

2) apply the consistency ($\rho^\sharp$);

3) and either discard the result $e$, add it to the set of solutions sols, or split it ($\oplus$).

The solver starts with the maximal element $\top^\sharp$ of $\mathcal{D}^\sharp$, which represents $\gamma(\top^\sharp) = \hat{D}$.

This algorithm corresponds to algorithm 2.1 in which the abstract domain $E$ defined in CP is replaced by the abstract domain $\mathcal{D}^\sharp$.

The termination is ensured by the following proposition:

PROPOSITION 5.1.– If $\tau$ and $\oplus$ are compatible, algorithm 5.1 terminates.

PROOF 5.1.– We show that the search tree is finite. Suppose that the search tree is infinite. Its width is finite by definition 5.5, and there would exist an infinite branch (König's lemma), which would contradict definition 5.7.

The search tree being finite, algorithm 5.1 terminates. □

PROPOSITION 5.2.– Algorithm 5.1 is correct.

PROOF 5.2.– At each iteration, $\bigcup\{\gamma(x) \mid x \in \text{toExplore} \cup \text{sols}\}$ is an overapproximation of the set of solutions, because the consistency $\rho^\sharp$ is an abstraction of the concrete semantics $\rho$ of the constraints and the splitting operator $\oplus$ is an abstraction of the identity. We note that abstract elements in sols are consistent and either contain only solutions or are smaller than $r$. The algorithm terminates when toExplore is empty, at which point sols overapproximates the set of solutions with consistent elements that contain only solutions or are smaller than $r$. To compute the exact set of solutions in the discrete case, it is sufficient to choose $r = 1$.

Thus, algorithm 5.1 computes an overapproximation or an exact approximation in the discrete case of the solutions set. □

**Algorithm 5.1.** *Our generic abstract solver. For a color version of the algorithm, see www.iste.co.uk/pelleau/ abstractdomains.zip*

list of abstract domains sols $\leftarrow \emptyset$ /* *stores the abstract solutions* */
queue of abstract domains toExplore $\leftarrow \emptyset$ /* *stores the abstract elements to explore* */
abstract domain $e \in \mathcal{D}^\sharp$

**push** $\top^\sharp$ in toExplore /* *initialization with the abstract search space:* $\gamma(\top^\sharp) = \hat{D}$ */

**while** toExplore $\neq \emptyset$ **do**
  $e \leftarrow$ **pop**(toExplore)
  $e \leftarrow \rho^\sharp(e)$
  **if** e $\neq \emptyset$ **then**
    **if** $\tau(e) \leq r$ **or** isSol(e) **then** /* *isSol(e) returns* ***true*** *if* e *contains only solutions* */
      sols $\leftarrow$ sols $\cup$ $e$
    **else**
      **push** $\oplus(e)$ in toExplore
    **end if**
  **end if**
**end while**

The solving process presented in algorithm 5.1 uses a queue data structure, and splits the oldest abstract element first. More clever choosing strategies, such as splitting the largest element for $\tau$, can be defined. As the two previous propositions do not depend on the choice operator, the algorithm remains correct and terminates for any strategy.

Similarly to local iterations in AI, our solver performs decreasing abstract iterations. More precisely, toExplore $\cup$ sols is decreasing for $\sqsubseteq^\sharp_\mathcal{E}$ in the disjunctive completion domain $\mathcal{E}^\sharp$ at each iteration of the loop. Indeed, $\rho^\sharp$ is reductive in $\mathcal{D}^\sharp$, $\rho^\sharp(e) \sqsubseteq^\sharp e$; moreover, the element returned by the splitting operator is included in the starting element

$\oplus(e) \sqsubseteq^{\sharp}_{\mathcal{E}} \{e\}$. However, our solver differs from classic AI in two ways. First, there is no splitting operator in AI, new components in a disjunctive completion are generally added only at control-flow joins. For instance, when analyzing a conditional, two elements are created: one verifying the condition and the other not satisfying it. The first element is used to analyze the instructions in the block and then the other to analyze the *else* block. However, once leaving the conditional, the abstract union $\sqcup^{\sharp}$ of the two elements is performed, and the analysis continues with one element. Second, the solving iteration strategy is far more elaborated than in AI. The use of a narrowing is replaced with a data structure that maintains an ordered list of abstract elements and a splitting strategy that performs a refinement process and ensures its termination. Actually, more complex strategies than the simple one we presented here exist in the CP literature. One example is the AC-5 algorithm [HEN 92] where each time the domain of a variable changes, and the variable decides which constraints need to be propagated. The design of efficient propagation algorithms is an active research area in CP [SCH 01].

## 5.2. The AbSolute solver

We have implemented a prototype abstract solver, from the ideas presented previously, to demonstrate the feasibility of our approach. We describe its main features and present experimental results.

### 5.2.1. *Implementation*

Our prototype solver, called AbSolute, is implemented in OCaml. It uses Apron, a library of numeric abstract domains intended primarily for static analysis [JEA 09]. Several abstract domains are implemented in Apron, such as intervals, octagons and polyhedra. Moreover, Apron offers an uniform application programming interface (API) that hides abstract domains internal algorithms. Thus, a function can be generically implemented, without knowing which abstract domain is going to be used. For instance, the consistency can be generically implemented for any abstract domain, and does not need to be specifically implemented for each abstract domain.

Another important feature in Apron is the fact that abstract domains are defined for both integer and real variables. In other words, an abstract domain can be defined for a set of variables that only contains integer or real variables, or that contains both integer and real variables.

Finally, Apron provides a language of arithmetic constraints sufficient to express many CSPs: equalities and inequalities over numeric expressions, including operators such as $+$, $-$, $\times$, $/$, $\sqrt{}$, power, modulo and rounding to integers.

For all these reasons, we choose Apron and take advantage of its abstract domains, its uniform API, its management of integer and real variables and nonlinear constraints.

#### 5.2.1.1. *Problem modelization*

In CP, a problem is formalized under the form of a CSP. To represent a CSP in AbSolute, the variables are stored in an environment composed of two tables: one for the integer variables and the other for the real variables. The constraints that can be expressed in Apron are stored in a table. Since the domain of the variables is a notion that does not exist in AI, they are translated into linear constraints and stored in the constraints table.

EXAMPLE 5.15 (Problem modelization in AbSolute).– Consider the CSP on integer variables $v_1, v_2$ with domains $D_1 = D_2 = [\![1, 5]\!]$ and the real variable $v_3$ with domain $D_3 = [-10, 10]$, and with the constraints:

$$C_1\colon 6v_1 + 4v_2 - 3v_3 = 0$$
$$C_2\colon v_1 \times v_2 \geq 3.5$$

It is translated in AbSolute as:

```
let v1 = Var.of_string "v1";;
let v2 = Var.of_string "v2";;
let v3 = Var.of_string "v3";;
```

```
let csp =
  let vars = Environment.make [|v1 ; v2|] [|v3|] in
  let doms = Parser.lincons1_of_lstring vars
  ["v1 ≥ 1";"v1 ≤ 5";"v2 ≥ 1"; "v2 ≤ 5";"v3 ≥ −10";"v3 ≤ 10"]
  in
  let cons = Parser.tcons1_of_lstring vars ["6 ∗ v1 − 4 ∗ v2 −
  3 ∗ v3 = 0"; "v1 ∗ v2 ≥ 3.5"] in
(vars, doms, cons);;
```

The first three lines correspond to the creation of the variables. They are created using the function `Var.of_string` that takes a parameter, the string corresponding to the name of the variable. Then, the CSP is created, the set of variables (`vars`) is created using the function `Environment.make` that takes as first argument, the set of integer variables (here $\{v_1, v_2\}$), and as second argument, the set of real variables ($\{v_3\}$). The domains are created with function `Parser.lincons1_of_lstring` which transforms a table of string into a table of linear constraints on variables `vars`. Finally, the conjunction of constraints is created using the function `Parser.tcons1_of_lstring` which transforms a table of string into a table of constraints (that can be nonlinear) on variables `vars`.

5.2.1.2. *Abstraction*

An abstract domain $\mathcal{D}^\sharp$ is chosen and initialized with the domains in the CSP. It is created using a "manager". Each abstract domain has its own manager. Then, the transfer function of the constraints corresponding to the domains is called to initialize the abstract domain to the initial search space.

EXAMPLE 5.16 (Creating an abstract domain in AbSolute).– We continue example 5.15. To create an abstract domain from the domains, we write in AbSolute:

```
let abs = Abstract1.of_lincons_array man vars doms;;
```

where `man` corresponds to the type of abstract domain that we want to create. For instance, we will use `Oct.manager_alloc()` for the octagons. Function `Abstract1.of_lincons_array` directly creates an abstract domains for variables `vars` with the table of linear constraints `doms`.

Finally, $\mathcal{D}^\sharp$ represents an abstract domain in AI (such as the octagons) and a conjunction of linear constraints corresponding to the domains in CP ($v_1 \geq 1, v_1 \leq 5$...).

5.2.1.3. *Consistency*

The test transfer function naturally provides propagators for the constraints. Internally, each domain implements its own algorithm to handle tests, including sophisticated methods to handle nonlinear constraints. For instance, in the intervals abstract domain, the algorithm HC4-Revise is implemented to better propagate the nonlinear constraints. For all the abstract domains, if no method is able to propagate nonlinear constraints, they are linearized using the algorithm described in [MIN 04]. This algorithm replaces in each nonlinear term a subset of variables by the interval of possible values for these variables. We thus have a *quasi-linear* constraint that is a constraint with intervals for coefficients. Then, each interval $[a, b]$ is replaced by the value in the middle $(a + b)/2$, and the interval $[(a - b)/2, (b - a)/2]$ is added to the constants.

EXAMPLE 5.17 (Linearization).– Still considering the CSP in example 5.15, constraint $C_2$ can be transformed into the following quasi-linear constraints:

$C_{2.1}$: $[\![1, 5]\!]v_2 \geq 3.5$
$C_{2.2}$: $[\![1, 5]\!]v_1 \geq 3.5$

Constraint $C_{2.1}$ corresponds to constraint $C_2$ in which variable $v_1$ has been replaced by its domain $[\![1, 5]\!]$. Similarly, constraint $C_{2.2}$ corresponds to constraint $C_2$ in which variable $v_2$ has been replaced by its domain $[\![1, 5]\!]$. Finally, the intervals are replaced by their median value and the interval $[\![-2, 2]\!]$ is added to the constants. We obtain the following linear constraints:

$C'_{2.1}$: $3v_2 + [\![-2, 2]\!] \geq 3.5$
$C'_{2.2}$: $3v_1 + [\![-2, 2]\!] \geq 3.5$

Other linearization algorithms can be considered, such as one algorithm proposed in [BOR 05] which, in the last step, replaced the intervals by either the upper or the lower limit according to the constraint and the starting interval.

In order to simulate the propagation loop, our solver performs local iterations until either a fixpoint or a maximum number of iterations is reached. This maximal number is set to 3 to avoid slow convergences and ensure fast solving.

EXAMPLE 5.18 (Slow convergence).– Consider the following CSP with two real variables $v_1$ and $v_2$ taking their values in domains $D_1 = D_2 = [-4, 4]$ and with the constraints:

$C_1$: $v_1 = v_2$
$C_2$: $v_1 = \frac{1}{2}v_2$

This problem has a unique solution $v_1 = v_2 = 0$, and the consistency has a very slow convergence. Indeed, by iteratively applying the propagators, the fixpoint is never reached. The propagator of constraint $C_2$ reduces the domain of $v_2$ to $[-2, 2]$, then the propagator of constraint $C_1$ reduces the domain of $v_1$ to $[-2, 2]$, then the propagator of constraint $C_2$ $v_2$ to $[-1, 1]$, and so on. The propagators reduce the domains of the variables by half at each call. The first three iterations are illustrated in Figure 5.1.

REMARK 5.3.– To avoid slow convergences, by default, AbSolute performs only three iterations in the consistency. In CP solvers, such as Ibex, the consistency is stopped if the iteration has reduced the domains of less than 10%.

REMARK 5.4.– In solvers in CP, only constraints containing at least one variable whose domain has been modified during the previous iteration are propagated. However, for simplicity, our solver propagates all the constraints at each step of the local iteration.

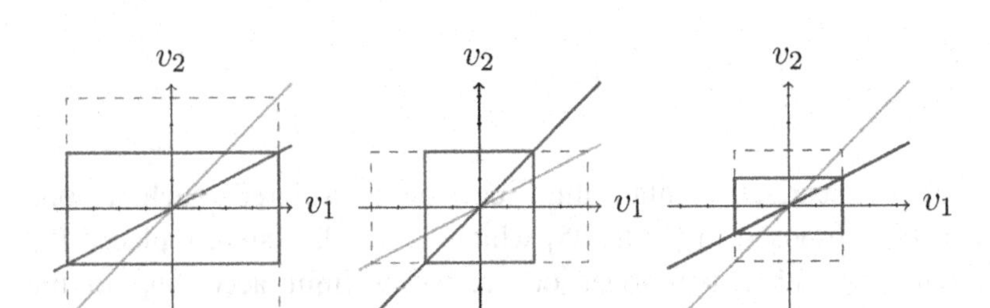


**Figure 5.1.** *Example of the three first iterations in the computation of the consistency with a slow convergence*

EXAMPLE 5.19 (Call to the consistency in AbSolute).– Continuing example 5.16. Once the abstract domain is created, the consistency can be called, and it is performed in AbSolute by the following instruction:

```
let abs = consistency man abs cons max_iter;;
```

where `max_iter` is the maximal number of iterations that can be performed and whose default value is 3. The type of the abstract domain `man` is mandatory so that Apron can execute the corresponding transfer functions.

#### 5.2.1.4. *Splitting operator*

Currently, our solver only splits along a single variable at a time. The splitting operator splits the largest domain in two, even for relational domains and integer variables. The splitting operator uses the smallest box containing the abstract element, computes the largest dimension and cuts it in two.

Figure 5.2(a) gives an example of the splitting operator applied to an octagon. The smallest box containing the octagon is first computed. Then, the variable $v_1$ is split along the dashed line. For comparison, Figure 5.2(b) illustrates the split performed by the octagonal splitting operator as given in definition 3.6.

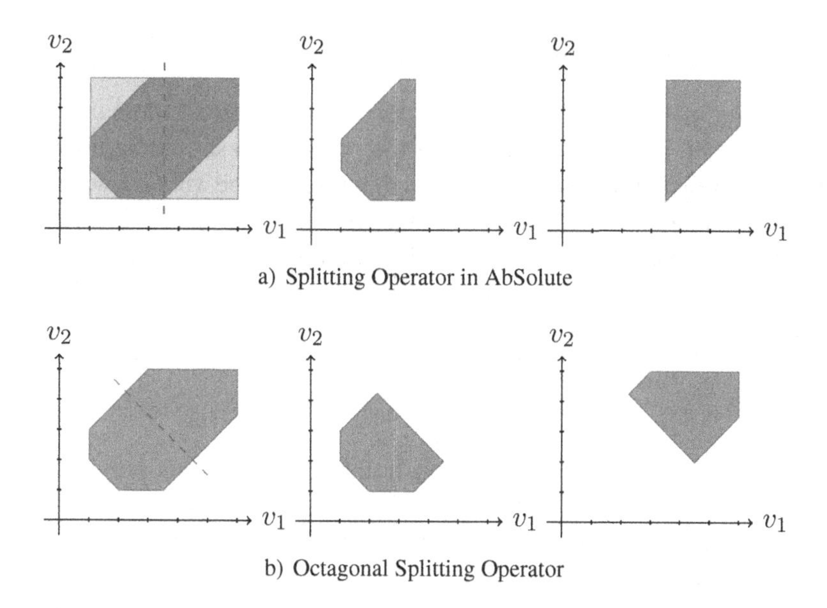


**Figure 5.2.** *Comparison between the naive splitting operator developed in AbSolute applied to an octagon, and the octagonal splitting operator given definition 3.6*

EXAMPLE 5.20 (Call to the splitting operator in AbSolute).– Continuing example 5.19. Once the abstract consistent domain computed, the splitting operator can be called, and it is performed in AbSolute using the following instruction:

```
let list_abs = split man abs vars;;
```

where the splitting operator `split` returns a list of abstract domains whose abstract union is equivalent to the starting abstract element `abs`.

Compared to most CP solvers, this splitting strategy is very basic, and depends neither on the abstract domain, nor on the type of variables (Figure 5.2(a)). Clever strategies from the CP literature must be consider in future works, such as the one defined in section 1.2.5.

5.2.1.5. *Polyhedra particular case*

In the particular case of the polyhedron abstract domain, we realized that, in practice, the size of the polyhedron should be limited. Indeed, the solving process with polyhedron may be very slow, since the consistency is not always reductive, and can add new linear expressions that are not always relevant. For instance, consider the polyhedron on variables $v_1$ and $v_2$ composed of the following linear expressions:

$$\{v_1 \geq 1, v_1 \leq 5, v_2 \geq 1, v_2 \leq 5\}$$

The consistency for a set of constraints applied to this polyhedron may want to add to the polyhedron the linear expression $v_2 - 5v_1 \leq 0$. This linear expression is always true for any point of the polyhedron. Thus, it is considered irrelevant. The figure below illustrates this example, the polyhedron is the square and the new linear expression corresponds to the shaded area. We can clearly see that any point of the polyhedron satisfies the linear expression since the shaded area totally covers the polyhedron.

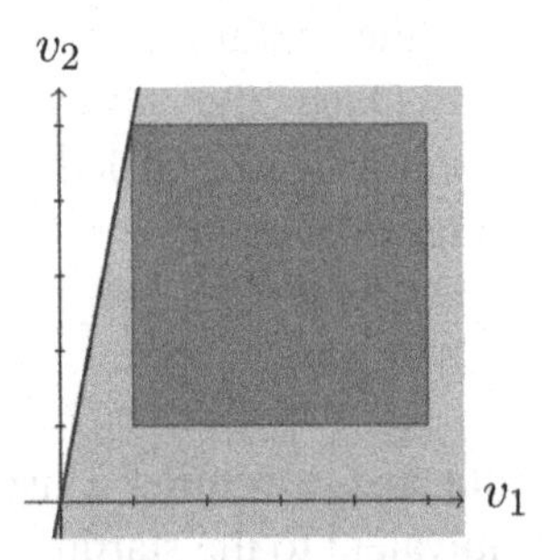


**Figure 5.3.** *Example of a polyhedron (square) that satisfies the linear expression (shaded area). Adding this linear expression to the polyhedron does not add any new information*

Finally, in the particular case of the polyhedron abstract domain, only the splitting operator is reductive. We thus decided to limit the size of the polyhedron by limiting its number of facets. Rather than giving a fixed number of facets, we chose to limit the value for the coefficient of the polyhedron linear expressions. By doing so, we do not need to verify after each split or consistency that the polyhedron still has the good number of facets. We only need to verify the coefficients of the newly added linear expressions.

We arbitrarily set to 20 this maximum value for the coefficient of the polyhedron linear expressions. We did not have the time to do exhaustive experiments to determine, if it exists, the best possible value for this limit. Moreover, on the test we did performed, we can see that the chosen value is not limiting enough.

#### 5.2.1.6. *Solving*

We have implemented two versions of the solving process. The first version corresponds to the solving method given by algorithm 5.1. The second one stops as soon as a first solution is found. In both versions, a queue is used to maintain the set of abstract elements to explore. No choice operator is implemented for the moment, and the abstract elements are visited following the order given by the queue. Both methods use the abstract domain and the value $r$ as parameters. More importantly, both methods do not depend on the problem to solve, and thus do not depend on the variable type. Therefore, they can both be used to solve problems containing both integer and real variables.

### 5.2.2. *Experimental results*

We have run AbSolute on two classes of problems: first, on continuous problems to compare AbSolute efficiency with state-of-the-art CP solvers; second, on mixed problems, that these CP solvers cannot handle while our abstract solver can.

#### 5.2.2.1. *Continuous solving*

We use problems from the Coconut benchmark[1], a standard CP benchmark with only real variables. We compare the results obtained

1 Available at www.mat.univie.ac.at/neum/glopt/coconut/.

with AbSolute to the ones obtained with a standard interval-based CP continuous solver called Ibex[2]. Additionally, we compare AbSolute to our extension of Ibex to octagons (as presented in section 4.4.1), which allows comparing the choice of domain (intervals vs. octagons) independently from the choice of solver algorithm (classic CP solver vs. our AI-based solver). Tables 5.2 and 5.1 show the run time in seconds to find all the solutions or only the first solution of each problem.

| | | | *Intervals* | | *Octagons* | |
|---|---|---|---|---|---|---|
| *Name* | *# vars* | *ctr type* | *Ibex* | *AbSolute* | *Ibex* | *AbSolute* |
| b | 4 | = | **0.02** | 0.10 | 0.26 | **0.14** |
| nbody5.1 | 6 | = | **95.99** | 1,538.25 | **27.08** | $\geq$ 1h |
| ipp | 8 | = | **38.83** | 39.24 | **279.36** | 817.86 |
| brent-10 | 10 | = | **21.58** | 263.86 | **330.73** | $\geq$ 1h |
| KinematicPair | 2 | $\leq$ | 59.04 | **23.14** | 60.78 | **31.11** |
| biggsc4 | 4 | $\leq$ | 800.91 | **414.94** | 1,772.52 | **688.56** |
| o32 | 5 | $\leq$ | 27.36 | **22.66** | 40.74 | **33.17** |

**Table 5.1.** *Comparison of the CPU time in seconds to find all solutions with Ibex and AbSolute*

| | | | *Intervals* | | *Octagons* | |
|---|---|---|---|---|---|---|
| *Name* | *# vars* | *ctr type* | *Ibex* | *AbSolute* | *Ibex* | AbSolute |
| b | 4 | = | **0.009** | 0.018 | 0.053 | **0.048** |
| nbody5.1 | 6 | = | **32.85** | 708.47 | **0.027** | $\geq$ 1h |
| ipp | 8 | = | **0.66** | 9.64 | 19.28 | **1.46** |
| brent-10 | 10 | = | 7.96 | **4.57** | **0.617** | $\geq$ 1h |
| KinematicPair | 2 | $\leq$ | **0.013** | 0.018 | 0.016 | **0.011** |
| biggsc4 | 4 | $\leq$ | **0.011** | 0.022 | 0.096 | **0.029** |
| o32 | 5 | $\leq$ | **0.045** | 0.156 | **0.021** | 0.263 |

**Table 5.2.** *Comparison of the CPU time in seconds to find the first solution with Ibex and AbSolute*

On average, AbSolute is competitive with the traditional CP approach. More precisely, it is globally slower on problems with equalities, and faster on problems with inequalities. This difference of

2 Available at www.emn.fr/z-info/ibex/.

performance does not seem to be related to the type of constraints but rather on the following ratio: the number of constraints in which a variable appears over the total number of constraints. As mentioned earlier, at each iteration, all the constraints are propagated even those for which none of their variables have changed. This increases the computation time at each step and thus increases the overall time. For instance, in the problem `brent-10`, there are 10 variables, 10 constraints and each variable appears in at most three constraints. If only one variable has been modified, we will nevertheless propagate all 10 constraints, instead of three at most. This may explain the time-outs observed in problems `brent-10` and `nbody5.1` with AbSolute.

Moreover, in our solver, the propagation loop is stopped after three iterations, while, in the classic CP approach, the fixpoint is reached. The consistency in AbSolute may be less precise than the one used in Ibex. This may reduce the time spent during the propagation step but may increase the search phase.

These experimentations show that our prototype, which only features quite naive CP strategies, behaves reasonably well on a classic benchmark. Further studies will include a deeper analysis of the performances and improvements of AbSolute on its identified weaknesses (splitting strategy, propagation loop). Since it is called at each node of the search tree, the propagation loop is a key point in complete solving method. The efficiency of a solving method depends on the efficiency of its propagation loop.

#### 5.2.2.2. *Mixed discrete-continuous solving*

As CP solvers seldom handle mixed problems, no standard benchmark exists. We thus gathered problems from MinLPLib[3], a library of mixed optimization problems from the operational research community. These problems are not constraints satisfaction problems, but optimization problems, with constraints to satisfy and a function to minimize. We thus needed to turn them into constraints satisfaction

3 Available at www.gamsworld.org/minlp/minlplib.htm.

problems, following the approach in [BER 09]. We replaced each optimization criterion $\min f(x)$ with a constraint $|f(x) - \text{best\_known\_value}| \leq \epsilon$. We compared AbSolute to the mixed solving scheme from [BER 09], using the same $\epsilon$ and benchmarks, and found that they have similar run times (we do not provide a more detailed comparison as it would be meaningless due to the machine differences).

More interestingly, we observe that AbSolute can solve mixed problems in reasonable time and behaves better with intervals than with relational domains. A possible reason is that current propagations and heuristics are not able to fully use relational information available in octagons or polyhedra. In section 4.3.1, we suggest that a carefully designed splitting operator is key to efficient octagons. Future work will incorporate ideas from the octagonal CP solver into our solver and develop them further. However, AbSolute is able to naturally cope with mixed CP problems in a reasonable time, opening the way to new CP applications such as the repair and restauration of the electrical power system after a natural disaster [SIM 12]. The goal is to find a recovery plan that is scheduled and the route of each repair team in order to restore, as fast as possible, the power network after a disaster. Another application is for geometrical problems, an extension of the problem presented in [BEL 07], in these problems, the goal is to place objects of various shapes (polygons, algebraic varieties, etc.) so that no object overlaps another while minimizing the surface or the volume occupied.

| Name | # vars | | ctr type | $\mathcal{M}^\sharp$ | $\mathcal{O}^\sharp$ | $\mathcal{P}^\sharp$ |
|---|---|---|---|---|---|---|
| | int | real | | | | |
| gear4 | 4 | 2 | $=$ | **0.017** | 0.048 | 0.415 |
| st_miqp5 | 2 | 5 | $\leq$ | **2.636** | 3.636 | $\geq$ 1h |
| ex1263 | 72 | 20 | $=$ $\leq$ | **473.933** | $\geq$ 1h | $\geq$ 1h |
| antennes_4_3 | 6 | 2 | $\leq$ | **520.766** | 1562.335 | $\geq$ 1h |

**Table 5.3.** *Comparison of the central processing unit (CPU) time, in seconds, to solve mixed problems with AbSolute using different abstract domains*

| Name | # vars | | ctr type | $\mathcal{M}^\sharp$ | $\mathcal{O}^\sharp$ | $\mathcal{P}^\sharp$ |
|---|---|---|---|---|---|---|
| | int | real | | | | |
| gear4 | 4 | 2 | $=$ | **0.016** | 0.036 | 0.296 |
| st_miqp5 | 2 | 5 | $\leq$ | **0.672** | 1.152 | $\geq$ 1h |
| ex1263 | 72 | 20 | $=$ $\leq$ | **8.747** | $\geq$ 1h | $\geq$ 1h |
| antennes_4_3 | 6 | 2 | $\leq$ | **3.297** | 22.545 | $\geq$ 1h |

**Table 5.4.** *Comparison of the CPU time, in seconds, to find the first solution of mixed problems with AbSolute using different abstract domains*

## 5.3. Conclusion

In this chapter, we have explored some links between AI and CP, and used them to design a constraint satisfaction problem-solving scheme built entirely on abstract domains. The preliminary results obtained with our prototype are encouraging and open the way to the development of hybrid CP–AI solvers able to naturally handle mixed constraint problems.

In future work, we would like to improve our solver by adapting and integrating advanced methods from the CP literature. The areas of improvement include: split operators for abstract domains, specialized propagators, such as octagonal consistency or global constraints, and improvements to the propagation loop.

We built our solver on abstractions in a modular way so that existing and new methods can be combined together, as is the case for reduced products in AI. When analyzing a program, several abstract domains are generally used. In this case, the reduced product is used to communicate the information to abstract domains, which was gathered during the analysis from other abstract domains. Ultimately, each problem should automatically be solved in the abstract domains which best fit it, as is the case in AI. As a result, linear problems should be solved using polyhedra and second-degree polynomials with ellipsoids.

Another exciting development would be to use some methods form CP in a static analyzer in AI, such as the use of a split operator in disjunctive completion domains, and the ability of CP to refine an abstract element to achieve completeness.

# Conclusion and Perspectives

## C.1. Conclusion

In this book, we first studied and compared certain methods of constraint programming (CP) and abstract interpretation (AI). We are particularly interested in tools to overapproximate impossible or difficult domains to be computed exactly (concrete domain AI and solutions set CP). We relied on this study to abstract the concept of domains used in CP, in the sense that the chosen representation for domains becomes a parameter of the solver. This is equivalent to integrating abstract domains of the AI framework into CP. This allows new representations for domains to be used, including relational representations that capture some relationships between variables (for example, the polyhedron abstract domain integrates linear relationships). Furthermore, the abstract domains allowed us to define a unified solving method that does not depend on the types of variables and uniformly integrates the discrete and continuous solving methods of CP.

Secondly, we have defined and implemented the weakly relational abstract domain of octagons in CP for continuous variables. We adapted the lower closure operator existing in AI on octagons to design an optimal octagonal consistency. We then used the relationship between variables expressed by octagons to guide the search. Tests performed on a classical benchmark show that these heuristics are effective. This new solving method shows the importance of developing new representations for domains in CP, especially to integrate relational domains that are more expressive than the Cartesian product generally used in CP. However, to do this, it is necessary for each new domain to redefine appropriate splitting operators and consistencies, which constitutes an obstacle.

Yet, operators quite similar to the consistency have already been defined in AI for many abstract domains. So, we decided to do the opposite of the previous work, and instead of integrating the concepts of AI in CP, we expressed CP under the AI framework. This is done by considering the set of all the searched solutions as the concrete domain. We have redefined constraint solving with operators in AI. This allows us to use existing abstract domains directly. We implemented this method over Apron, a library of abstract domains. The prototype developed, AbSolute, which has not integrated CP exploration strategies yet, gave promising early results.

## C.2. Perspectives

This book opens up several research perspectives. First, in the short term, AbSolute must be developed to integrate more clever research techniques, such as heuristics for the variable/value choice, splitting operators or setting of the propagation loop (number and order of local iterations). For relational domains (octagons and polyhedra) in particular, work on octagons suggests that the key to effectiveness is to use existing relationships between variables to guide the search. For mixed domains, the current method can be improved by defining and implementing reduced products between real and integer domains. Moreover, AbSolute must also be tested on real applications, especially for mixed problems such as the problem of restoration and repair of the power grid after a natural disaster [SIM 12]. The results obtained from these applications can lead to new dedicated heuristics.

In the medium term, we should study the many existing abstract domains in AI, such as the interval polyhedra, polyhedra, zonotopes or ellipsoids, as part of the CP. This means defining associated consistencies and splitting operators. In particular, for polyhedra, we should find an effective approximation of nonlinear constraints, which requires a thorough study of existing linearization or quasi-linearization algorithms.

In the long term, we finally plan to exploit the link between underapproximations in CP and the fixed point computed in AI. In AI, efficient analyzers are mainly based on the computation of the widening operator. We can exploit the important work on the widening operator to adapt it to a solving method for inner approximation in CP.

# Bibliography

[ABD 96] ABDALLAH C., DORATO P., LISKA R. *et al.*, "Applications of quantifier elimination theory to control theory", *Proceedings of the 4th IEEE Mediterranean Symposium on New Directions in Control and Automation*, 1996.

[ALB 05] ALBERGANTI M., "L'avion qui "bat des ailes" a fédéré de nombreux chercheurs", *Le Monde*, vol. 18741, p. 18, 2005.

[ALP 93] ALPUENTE M., FALASCHI M., RAMIS M.J. *et al.*, "Narrowing approximations as an optimization for equational logic Programs", *Proceedings of the 5th International Symposium on Programming Language Implementation and Logic Programming (PLILP)*, Lecture Notes in Computer Science, Springer-Verlag, pp. 391–409, 1993.

[ANS 09] ANSÓTEGUI C., SELLMANN M., TIERNEY K., "A gender-based genetic algorithm for the automatic configuration of algorithms", *Proceedings of the 15th International Conference on Principles and Practice of Constraint Programming (CP)*, Lecture Notes in Computer Science, Springer-Verlag, vol. 5732, pp. 142–157, 2009.

[APT 99] APT K.R., "The essence of constraint propagation", *Theoretical Computer Science*, vol. 221, 1999.

[APT 03] APT K.R., *Principles of Constraint Programming*, Cambridge University Press, New York, NY, USA, 2003.

[APT 07] APT K.R., WALLACE M., *Constraint Logic Programming Using Eclipse*, Cambridge University Press, New York, NY, USA, 2007.

[ARA 10] ARAYA I., TROMBETTONI G., NEVEU B., "Exploiting monotonicity in interval constraint propagation", *Proceedings of the 24th AAAI Conference on Artificial Intelligence (AAAI)*, 2010.

[ARA 12] ARAYA I., NEVEU B., TROMBETTONI G., "An interval extension based on occurrence grouping", *Computing*, vol. 94, pp. 173–188, 2012.

[ARB 09] ARBELAEZ A., HAMADI Y., SEBAG M., "Online heuristic selection in constraint programming", *Proceedings of the 4th International Symposium on Combinatorial Search (SoCS)*, 2009.

[BAG 05a] BAGNARA R., HILL P.M., MAZZI E. *et al.*, "Widening operators for weakly-relational numeric abstractions", *Proceedings of the 12th International Static Analysis Symposium (SAS)*, Lecture Notes in Computer Science, Springer, vol. 3672, pp. 3–18, 2005.

[BAG 05b] BAGNARA R., HILL P.M., RICCI E. *et al.*, "Precise widening operators for convex polyhedra", *Science of Computer Programming – Special issue: Static analysis symposium (SAS 2003)*, vol. 58, no. 1–2, pp. 28–56, 2005.

[BAG 06] BAGNARA R., HILL P.M., ZAFFANELLA E., "Widening operators for powerset domains", *International Journal on Software Tools for Technology Transfer*, vol. 8, no. 4, pp. 449–466, 2006.

[BAG 09] BAGNARA R., HILL P.M., ZAFFANELLA E., "Weakly-relational shapes for numeric abstractions: improved algorithms and proofs of correctness", *Formal Methods in System Design*, vol. 35, no. 3, pp. 279–323, 2009.

[BEE 06] VAN BEEK P., "Backtracking search algorithms", ROSSI F., VAN BEEK P., WALSH T. (eds.), *Handbook of Constraint Programming*, Elsevier, 2006.

[BEL 07] BELDICEANU N., CARLSSON M., PODER E. *et al.*, "A generic geometrical constraint kernel in space and time for handling polymorphic *k*-dimensional objects", *Proceedings of the 13th International Conference on Principles and Practice of Constraint Programming (CP)*, Lecture Notes in Computer Science, Springer, vol. 4741, pp. 180–194, 2007.

[BEL 10] BELDICEANU N., CARLSSON M., RAMPON J.-X., Global Constraint Catalog, 2nd Edition, Report no. T2010:07, The Swedish Institute of Computer Science, 2010.

[BEN 94] BENHAMOU F., MCALLESTER D.A., VAN HENTENRYCK P., "CLP(intervals) revisited", *Proceedings of the 1994 International Symposium on Logic programming (ILPS)*, MIT Press, pp. 124–138, 1994.

[BEN 95] BENHAMOU F., "Interval constraint logic programming", *Constraint Programming: Basics and Trends*, Lecture Notes in Computer Science, Springer Berlin/Heidelberg, vol. 910, pp. 1–21, 1995.

[BEN 96] BENHAMOU F., "Heterogeneous constraint solvings", *Proceedings of the 5th International Conference on Algebraic and Logic Programming*, pp. 62–76, 1996.

[BEN 97a] BENHAMOU F., GOUALARD F., GRANVILLIERS L., "Programming with the DecLIC Language", *Proceedings of the 2nd International Workshop on Interval Constraints*, 1997.

[BEN 97b] BENHAMOU F., OLDER W.J., "Applying interval arithmetic to real, integer and Boolean constraints", *Journal of Logic Programming*, vol. 32, no. 1, pp. 1–24, 1997.

[BEN 99] BENHAMOU F., GOUALARD F., GRANVILLIERS L. *et al.*, "Revisiting hull and box consistency", *Proceedings of the 16th International Conference on Logic Programming*, pp. 230–244, 1999.

[BEN 04] BENHAMOU F., GOUALARD F., LANGUENOU É. *et al.*, "Interval constraint solving for camera control and motion planning", *ACM Transactions on Computational Logic*, vol. 5, no. 4, pp. 732–767, 2004.

[BER 09] BERGER N., GRANVILLIERS L., "Some interval approximation techniques for MINLP", *Proceedings of the The 8th Symposium on Abstraction, Reformulation and Approximation (SARA)*, 2009.

[BER 10] BERTRANE J., COUSOT P., COUSOT R. *et al.*, "Static analysis and verification of aerospace software by abstract interpretation", *AIAA Infotech@Aerospace*, Atlanta, Georgia, American Institute of Aeronautics and Astronautics, 20–22 April 2010.

[BES 94] BESSIÈRE C., "Arc-consistency and arc-consistency again", *Artificial Intelligence*, vol. 65, no. 1, pp. 179–190, 1994.

[BES 96] BESSIÈRE C., RÉGIN J.-C., "MAC and combined heuristics: two reasons to forsake FC (and CBJ?) on hard problems", *Proceedings of the Second International Conference on Principles and Practice of Constraint Programming*, Lecture Notes in Computer Science, Springer, vol. 1118, 1996.

[BES 99] BESSIÈRE C., FREUDER E.C., RÉGIN J.-C., "Using constraint metaknowledge to reduce arc consistency computation", *Artificial Intelligence*, vol. 107, no. 1, pp. 125–148, 1999.

[BES 01] BESSIÈRE C., RÉGIN J.-C., "Refining the basic constraint propagation algorithm", *Proceedings of the 17th International Joint Conference on Artificial Intelligence (IJCAI)*, Morgan Kaufmann, pp. 309–315, 2001.

[BES 03] BESSIÈRE C., VAN HENTENRYCK P., "To be or not to be ... a global constraint", *Proceedings of the 9th International Conference on Principles and Practice of Constraint Programming (CP)*, Lecture Notes in Computer Science, Springer, vol. 2833, pp. 789–794, 2003.

[BES 04] BESSIÈRE C., HEBRARD E., HNICH B. *et al.*, "The Complexity of Global Constraints", *Proceedings of the 19th Conference on Artificial Intelligence (AAAI)*, pp. 112–117, 2004.

[BES 06] BESSIÈRE C., "Constraint propagation", ROSSI F., VAN BEEK P., WALSH T. (eds.), *Handbook of Constraint Programming*, Elsevier, 2006.

[BES 11] BESSIERE C., CARDON S., DEBRUYNE R. *et al.*, "Efficient algorithms for singleton arc consistency", *Constraints*, vol. 16, no. 1, pp. 25–53, 2011.

[BOR 05] BORRADAILE G., VAN HENTENRYCK P., "Safe and tight linear estimators for global optimization", *Mathematical Programming*, vol. 102, pp. 495–517, 2005.

[BOU 92] BOURDONCLE F., "Abstract interpreting by dynamic partitioning", *Journal of Functional Programming*, vol. 2, pp. 407–435, 1992.

[BOU 04] BOUSSEMART F., HEMERY F., LECOUTRE C. *et al.*, "Boosting systematic search by weighting constraints", *Proceedings of the 16th Eureopean Conference on Artificial Intelligence (ECAI)*, IOS Press, pp. 146–150, 2004.

[BRÉ 79] BRÉLAZ D., "New methods to color the vertices of a graph", *Communications of the ACM*, vol. 22, no. 4, pp. 251–256, 1979.

[CHA 09a] CHABERT G., JAULIN L., "Contractor programming", *Artificial Intelligence*, vol. 173, pp. 1079–1100, 2009.

[CHA 09b] CHABERT G., JAULIN L., LORCA X., "A constraint on the number of distinct vectors with application to localization", *Proceedings of the 15th International Conference on Principles and Practice of Constraint Programming (CP)*, Berlin, Heidelberg, Springer-Verlag, pp. 196–210, 2009.

[CHE 09] CHEN L., MINÉ A., WANG J. *et al.*, "Interval polyhedra: an abstract domain to infer interval linear relationships", *Proceedings of the 16th International Static Analysis Symposium (SAS)*, pp. 309–325, 2009.

[CHI 08] CHI K., JIANG X., HORIGUCHI S. *et al.*, "Topology design of network-coding-based multicast networks", *IEEE Transactions on Parallel and Distributed Systems*, vol. 19, no. 5, pp. 627–640, 2008.

[CHO 06] CHOI C.W., HARVEY W., LEE J.H.-M. *et al.*, "Finite domain bounds consistency revisited", *Proceedings of the 19th Australian Joint Conference on Artificial Intelligence: Advances in Artificial Intelligence (AI)*, Lecture Notes in Computer Science, Springer-Verlag, vol. 4304, pp. 49–58, 2006.

[CHO 10] CHOCO TEAM, Choco: an Open Source Java Constraint Programming Library, Research report no. 10-02-INFO, Ecole des Mines de Nantes, 2010.

[CHR 06] CHRISTIE M., NORMAND J.-M., TRUCHET C., "Calcul d'approximations intérieures pour la résolution de Max-NCSP", *Proceedings des Deuxièmes Journées Francophones de Programmation par Contraintes (JFPC)*, 2006.

[CLA 03] CLARKE E., GRUMBERG O., JHA S. *et al.*, "Counterexample-guided abstraction refinement for symbolic model checking", *Journal of the ACM*, vol. 50, pp. 752–794, ACM, 2003.

[CLA 04] CLARISÓ R., CORTADELLA J., "The octahedron abstract domain", *Proceedings of the 11th International Static Analysis Symposium (SAS)*, pp. 312–327, 2004.

[COL 94] COLMERAUER A., Spécifications de prolog IV, Report, Faculté des Sciences de Luminy, Marseille, France, 1994.

[COL 99] COLLAVIZZA H., DELOBEL F., RUEHER M., "Extending consistent domains of numeric CSP", *Proceedings of the 16th International Joint Conference on Artificial Intelligence*, pp. 406–413, 1999.

[COL 07] COLLAVIZZA H., RUEHER M., "Exploring different constraint-based modelings for program verification", *Proceedings of the 13th International Conference on Principles and Practice of Constraint Programming (CP)*, Lecture Notes in Computer Science, Springer, vol. 4741, pp. 49–63, 2007.

[COR 08] CORTESI A., "Widening operators for abstract interpretation", CERONE A., GRUNER S. (eds.), *Proceedings of the 6th IEEE International Conference on Software Engineering and Formal Methods*, pp. 31–40, 2008.

[COR 11] CORTESI A., ZANIOLI M., "Widening and narrowing operators for abstract interpretation", *Computer Languages, Systems & Structures*, vol. 37, no. 1, pp. 24–42, 2011.

[COU 76] COUSOT P., COUSOT R., "Static determination of dynamic properties of programs", *Proceedings of the 2nd International Symposium on Programming*, pp. 106–130, 1976.

[COU 77a] COUSOT P., COUSOT R., "Abstract interpretation: a unified lattice model for static analysis of programs by construction or approximation of fixpoints", *Conference Record of the Fourth Annual ACM SIGPLAN-SIGACT Symposium on Principles of Programming Languages*, Los Angeles, California, ACM Press, New York, NY, pp. 238–252, 1977.

[COU 77b] COUSOT P., COUSOT R., "Static determination of dynamic properties of generalized type unions", *Proceedings of an ACM Conference on Language Design for Reliable Software*, New York, NY, USA, ACM, pp. 77–94, 1977.

[COU 78] COUSOT P., HALBWACHS N., "Automatic discovery of linear restraints among variables of a program", *Proceedings of the 5th ACM SIGACT-SIGPLAN Symposium on Principles of Programming Languages*, pp. 84–96, 1978.

[COU 79] COUSOT P., COUSOT R., "Systematic design of program analysis frameworks", *Proceedings of the 6th ACM SIGACT-SIGPLAN Symposium of Principles of Programming Languages*, pp. 269–282, 1979.

[COU 92a] COUSOT P., COUSOT R., "Abstract interpretation frameworks", *Journal of Logic and Computation*, Oxford University Press, Oxford, UK, vol. 2, no. 4, pp. 511–547, August 1992.

[COU 92b] COUSOT P., COUSOT R., "Comparing the galois connection and widening/narrowing approaches to abstract interpretation", BRUYNOOGHE M., WIRSING M. (eds.), *Proceedings of the 4th International Symposium on Programming Language Implementation and Logic Programming (PLILP)*, Lecture Notes in Computer Science, Springer, vol. 631, pp. 269–295, 1992.

[COU 07] COUSOT P., COUSOT R., FERET J. *et al.*, "Combination of abstractions in the ASTRÉE static analyzer", *Proceedings of the 11th Asian Computing Science Conference on Advances in Computer Science: Secure Software and Related Issues (ASIAN 2006)*, Lecture Notes in Computer Science, Springer-Verlag, vol. 4435, pp. 272–300, 2007.

[COU 11] COUSOT P., COUSOT R., MAUBORGNE L., "The reduced product of abstract domains and the combination of decision procedures", *Proceedings of the 14th International Conference on Fondations of Software Science and Computation Structures (FoSSaCS)*, Lecture Notes in Computer Science, Springer-Verlag, vol. 6604, pp. 456–472, 2011.

[COV 03] COVERITY, http://www.coverity.com/, 2003.

[COV 11] COVERITY, CERN Chooses Coverity to Ensure Accuracy of Large Hadron Collider Software, Available at http://www.coverity.com/press-releases/cern-chooses-coverity-to-ensure-accuracy-of-large-hadron-collider-software/, 2011.

[COV 12] COVERITY, NASA Jet Propulsion Laboratory Relies on Coverity to Ensure the Seamless Touchdown and Operation of the Curiosity Mars Rover, available at http://www.coverity.com/press-releases/nasa-jpl-relies-on-coverity-to-ensure-the-seamless-touchdown-and-operation-of-the-curiosity-mars-rover/, 2012.

[DAN 97] DANTZIG G.B., THAPA M.N., *Linear Programming 1: Introduction*, Springer-Verlag, 1997.

[DEC 87] DECHTER R., PEARL J., "Network-based heuristics for constraint-satisfaction problems", *Artificial Intelligence*, vol. 34, no. 1, pp. 1–38, 1987.

[DEC 89] DECHTER R., MEIRI I., PEARL J., "Temporal constraint networks", *Proceedings of the 1st International Conference on Principles of Knowledge Representation and Reasoning*, 1989.

[DEC 90] DECHTER R., "Enhancement schemes for constraint processing: backjumping, learning, and cutset decomposition", *Artificial Intelligence*, vol. 41, no. 3, pp. 273–312, 1990.

[DEC 03] DECHTER R., *Constraint Processing*, Elsevier Morgan Kaufmann, 2003.

[DIA 12] DIAZ D., ABREU S., CODOGNET P., "On the implementation of Gnu Prolog", *Theory and Practice of Logic Programming*, Cambridge University Press, vol. 12, pp. 253–282, 2012.

[D'SI 06] D'SILVA V., Widening for automata, PhD thesis, Zürich University, 2006.

[D'SI 12] D'SILVA V., HALLER L., KROENING D., "Satisfiability solvers are static analysers", *Proceedings of the 19th International Static Analysis Symposium (SAS)*, Lecture Notes in Computer Science, Springer, vol. 7460, 2012.

[FAG 11] FAGES J.-G., LORCA X., "Revisiting the tree Constraint", *Proceedings of the 17th International Conference on Principles and Practice of Constraint Programming (CP)*, Lecture Notes in Computer Science, Springer-Verlag, vol. 6876, pp. 271–285, 2011.

[FAG 14] FAGES J.-G., CHABERT G., PRUD'HOMME C., "Combining finite and continuous solvers", *Computing Research Repository (CoRR)*, vol. abs/1402.1361, 2014.

[FER 04] FERET J., "Static analysis of digital filters", SPRINGER (ed.), *European Symposium on Programming (ESOP)*, vol. 2986, pp. 33–48, 2004.

[FLO 62] FLOYD R., "Algorithm 97: Shortest path", *Communications of the ACM*, vol. 5, no. 6, 1962.

[FRE 78] FREUDER E.C., "Synthesizing constraint expressions", *Communications of the ACM*, vol. 21, no. 11, pp. 958–966, 1978.

[FRE 82] FREUDER E.C., "A sufficient condition for Backtrack-Free search", *Journal of the ACM (JACM)*, vol. 29, no. 1, pp. 24–32, 1982.

[FRE 97] FREUDER E.C., "In pursuit of the holy grail", *Constraints*, vol. 2, no. 1, pp. 57–61, 1997.

[FRO 95] FROST D., DECHTER R., "Look-ahead value ordering for constraint satisfaction problems", *Proceedings of the 14th International Joint Conference on Artificial Intelligence (IJCAI)*, Morgan Kaufmann Publishers Inc., pp. 572–578, 1995.

[GEE 92] GEELEN P.A., "Dual viewpoint heuristics for binary constraint satisfaction problems", *Proceedings of the 10th European Conference on Artificial Intelligence (ECAI)*, John Wiley & Sons, Inc., pp. 31–35, 1992.

[GEN 96] GENT I.P., MACINTYRE E., PROSSER P. *et al.*, "An empirical study of dynamic variable ordering heuristics for the constraint satisfaction problem", *Proceedings of the 2nd International Conference on Principles and Practice of Constraint Programming*, Lecture Notes in Computer Science, Springer, vol. 1118, pp. 179–193, 1996.

[GEN 06] GENT I.P., JEFFERSON C., MIGUEL I., "MINION: a fast, scalable, constraint solver", *Proceedings of 17th European Conference on Artificial Intelligence (ECAI )*, IOS Press, pp. 98–102, 2006.

[GEN 08] GENT I.P., MIGUEL I., NIGHTINGALE P., "Generalised arc consistency for the AllDifferent constraint: an empirical survey", *Artificial Intelligence*, vol. 172, no. 18, pp. 1973–2000, 2008.

[GIN 90] GINSBERG M.L., FRANK M., HALPIN M.P. *et al.*, "Search lessons learned from crossword puzzles", *Proceedings of the 8th National Conference on Artificial Intelligence (AAAI)*, AAAI Press, pp. 210–215, 1990.

[GOL 91] GOLDBERG D., "What every computer scientist should know about floating point arithmetic", *ACM Computing Surveys*, vol. 23, no. 1, pp. 5–48, 1991.

[GOL 10] GOLDSZTEJN A., GRANVILLIERS L., "A new framework for sharp and efficient resolution of NCSP with manifolds of solutions", *Constraints*, vol. 15, no. 2, pp. 190–212, 2010.

[GRA 92] GRANGER P., "Improving the results of static analyses of programs by local decreasing iterations", *Proceedings of the 12th Conference on Foundations of Software Technology and Theoretical Computer Science*, 1992.

[GRA 06] GRANVILLIERS L., BENHAMOU F., "RealPaver: an interval solver using constraint satisfaction techniques", *ACM Transactions on Mathematical Software*, vol. 32, no. 1, pp. 138–156, 2006.

[GRI 11] GRIMES D., HEBRARD E., "Models and strategies for variants of the job shop scheduling problem", *Proceedings of the 17th International Conference on Principles and Practice of Constraint Programming (CP)*, Lecture Notes in Computer Science, Springer-Verlag, vol. 6876, pp. 356–372, 2011.

[HAL 12] HALBWACHS N., HENRY J., "When the decreasing sequence fails", *Proceedings of the 19th International Static Analysis Symposium (SAS)*, Lecture Notes in Computer Science, Springer, vol. 7460, 2012.

[HAN 92] HANSEN E., *Global Optimization Using Interval Analysis*, Marcel Dekker, 1992.

[HAR 79] HARALICK R.M., ELLIOTT G.L., "Increasing tree search efficiency for constraint satisfaction problems", *Proceedings of the 6th International Joint Conference on Artificial Intelligence (IJCAI)*, Morgan Kaufmann Publishers Inc., pp. 356–364, 1979.

[HAV 09] HAVELUND K., GROCE A., SMITH M. *et al.*, Monitoring the execution of space craft flight software, available at http://compass.informatik.rwth-aachen.de/ws-slides/havelund.pdf, 2009.

[HEN 92] VAN HENTENRYCK P., DEVILLE Y., TENG C.-M., "A generic arc-consistency algorithm and its specializations", *Artificial Intelligence*, vol. 57, 1992.

[HEN 95] VAN HENTENRYCK P., SARASWAT V.A., DEVILLE Y., "Design, implementation, and evaluation of the constraint language cc(FD)", *Selected Papers from Constraint Programming: Basics and Trends*, Springer-Verlag, pp. 293–316, 1995.

[HEN 97] VAN HENTENRYCK P., MICHEL L., DEVILLE Y., *Numerica: A Modeling Language for Global Optimization*, MIT Press, 1997.

[HEN 05] VAN HENTENRYCK P., MICHEL L., *Constraint-Based Local Search*, MIT Press, 2005.

[HEN 08] VAN HENTENRYCK P., YIP J., GERVET C. *et al.*, "Bound consistency for binary length-lex set constraints", *Proceedings of the 23rd National Conference on Artificial intelligence (AAAI)*, AAAI Press, pp. 375–380, 2008.

[HER 11a] HERMENIER F., DEMASSEY S., LORCA X., "Bin repacking scheduling in virtualized datacenters", *Proceedings of the 17th International Conference on Principles and Practice of Constraint Programming (CP'11)*, Lecture Notes in Computer Science, Springer-Verlag, vol. 6876, pp. 27–41, 2011.

[HER 11b] HERVIEU A., BAUDRY B., GOTLIEB A., "PACOGEN: automatic generation of pairwise test configurations from feature models", *Proceedings of the 22nd International Symposium on Software Reliability Engineering*, pp. 120–129, 2011.

[HIC 97] HICKEY T., JU Q., Efficient implementation of interval arithmetic narrowing using IEEE arithmetic, Report, IEEE Arithmetic, Brandeis University CS Dept, 1997.

[HOE 04] HOEVE W., "A hyper-arc consistency algorithm for the soft Alldifferent constraint", *Proceedings of the 10th International Conference on Principles and Practice of Constraint Programming (CP)*, Lecture Notes in Computer Science, Springer, vol. 3258, pp. 679–689, 2004.

[JEA 09] JEANNET B., MINÉ A., "Apron: a library of numerical abstract domains for static analysis", *Proceedings of the 21th International Conference Computer Aided Verification (CAV)*, Lecture Notes in Computer Science, Springer, vol. 5643, pp. 661–667, June 2009.

[KAS 04] KASK K., DECHTER R., GOGATE V., "Counting-based look-ahead schemes for constraint satisfaction", *Proceedings of the 10th International Conference on Principles and Practice of Constraint Programming (CP)*, Lecture Notes in Computer Science, Springer, vol. 3258, pp. 317–331, 2004.

[KAT 03] KATRIEL I., THIEL S., "Fast bound consistency for the global cardinality constraint", *Proceedings of the 9th International Conference on Principles and Practice of Constraint Programming (CP')*, Lecture Notes in Computer Science, Springer Berlin / Heidelberg, vol. 2833, pp. 437–451, 2003.

[KEA 96] KEARFOTT R.B., *Rigorous Global Search: Continuous Problems*, Kluwer, 1996.

[KIN 69] KING J.C., A program verifier, PhD thesis, Carnegie Mellon University, Pittsburgh, PA, 1969.

[KRO 08] KROENING D., STRICHMAN O., *Decision Procedures*, Springer, 2008.

[KRO 14] KROENING D., TAUTSCHNIG M., "CBMC – C bounded model checker", *Proceedings of the 20th International Conference on Tools and Algorithms for the Construction and Analysis of Systems (TACAS)*, Lecture Notes in Computer Science, Springer-Verlag, vol. 8413, pp. 389–391, 2014.

[LAC 98] LACAN P., MONFORT J.N., RIBAL L.V.Q. *et al.*, "ARIANE 5 – the software reliability verification process", *Proceedings of the Conference on Data Systems in Aerospace*, 1998.

[LAZ 12] LAZAAR N., GOTLIEB A., LEBBAH Y., "A CP framework for testing CP", *Constraints*, vol. 17, no. 2, pp. 123–147, 2012.

[LEC 03] LECOUTRE C., BOUSSEMART F., HEMERY F., "Exploiting multidirectionality in coarse-grained arc consistency algorithms", *Proceedings of the 9th International Conference on Principles and Practice of Constraint Programming (CP)*, Lecture Notes in Computer Science, Springer, vol. 2833, pp. 480–494, 2003.

[LIO 96] LIONS J.-L., Ariane 501 Inquest Report, available at http://www.astrosurf.com/luxorion/astronautique-accident-ariane-v501 .htm, 1996.

[LÓP 03] LÓPEZ-ORTIZ A., QUIMPER C.-G. *et al.*, "A fast and simple algorithm for bounds consistency of the all different constraint", *Proceedings of the 18th International Joint Conference on Artificial Intelligence (IJCAI)*, Morgan Kaufmann Publishers Inc., pp. 245–250, 2003.

[MAC 77a] MACKWORTH A.K., "Consistency in Networks of Relations", *Artificial Intelligence*, vol. 8, no. 1, pp. 99–118, 1977.

[MAC 77b] MACKWORTH A.K., "On reading sketch maps", *Proceedings of the 5th International Joint Conference on Artificial Intelligence*, pp. 598–606, 1977.

[MAC 85] MACKWORTH A.K., FREUDER E.C., "The complexity of some polynomial network consistency algorithms for constraint satisfaction problems", *Artificial Intelligence*, vol. 25, no. 1, pp. 65–74, 1985.

[MEH 00] MEHLHORN K., THIEL S., "Faster algorithms for bound-consistency of the sortedness and the Alldifferent constraint", *Proceedings of the 6th International Conference on Principles and Practice of Constraint Programming (CP)*, Lecture Notes in Computer Science, Springer, vol. 1894, pp. 306–319, 2000.

[MEN 83] MENASCHE M., BERTHOMIEU B., "Time petri nets for analyzing and verifying time dependent communication protocols", *Protocol Specification, Testing, and Verification*, 1983.

[MIN 04] MINÉ A., Domaines numériques abstraits faiblement relationnels, PhD thesis, École Normale Supérieure, December 2004.

[MIN 06] MINÉ A., "The octagon abstract domain", *Higher-Order and Symbolic Computation*, vol. 19, no. 1, pp. 31–100, Springer, 2006.

[MIN 12] MINÉ A., "Abstract domains for bit-level machine integer and floating-point operations", *Proceedings of The 4th International Workshop on Invariant Generation (WING)*, EpiC, EasyChair, p. 16, 2012.

[MIS 92] GAO Report: Patriot Missile Defense, available at http://www.fas.org/spp/starwars/gao/im92026.htm, 1992.

[MOH 86] MOHR R., HENDERSON T.C., "Arc and path consistency revisited", *Artificial Intelligence*, vol. 28, no. 2, pp. 225–233, 1986.

[MOH 88] MOHR R., MASINI G., "Good old discrete relaxation", *Proceedings of the 8th European Conference on Artificial Intelligence*, pp. 651–656, 1988.

[MON 74] MONTANARI U., "Networks of constraints: fundamental properties and applications to picture processing", *Information Science*, vol. 7, no. 2, pp. 95–132, 1974.

[MON 09] MONNIAUX D., "A minimalistic look at widening operators", *Higher Order and Symbolic Computation*, vol. 22, no. 2, pp. 145–154, 2009.

[MOO 66] MOORE R.E., *Interval Analysis*, Prentice-Hall, Englewood Cliffs N.J., 1966.

[MOR 93] MORENO-NAVARRO J.J., KUCHEN H., NO CARBALLO J.M. *et al.*, "Efficient lazy narrowing using demandedness analysis", *Proceedings of the 5th International Symposium on Programming Language Implementation and Logic Programming (PLILP)*, Lecture Notes in Computer Science, Springer-Verlag, pp. 167–183, 1993.

[PAC 01] PACHET F., ROY P., "Musical harmonization with constraints: a survey", *Constraints*, vol. 6, no. 1, pp. 7–19, 2001.

[PEL 09] PELLEAU M., VAN HENTENRYCK P., TRUCHET C., "Sonet network design problems", *Proceedings of the 6th International Workshop on Local Search Techniques in Constraint Satisfaction*, pp. 81–95, 2009.

[PEL 11] PELLEAU M., TRUCHET C., BENHAMOU F., "Octagonal domains for continuous constraints", *Proceedings of the 17th International Conference on Principles and Practice of Constraint Programming (CP)*, Lecture Notes in Computer Science, Springer-Verlag, vol. 6876, pp. 706–720, 2011.

[PEL 13] PELLEAU M., MINÉ A., TRUCHET C. *et al.*, "A constraint solver based on abstract domains", *Proceedings of the 14th International Conference on Verification, Model Checking, and Abstract Interpretation (VMCAI)*, Lecture Notes in Computer Science, Springer, Berlin Heidelberg, vol. 7737, pp. 434–454, 2013.

[PEL 14] PELLEAU M., TRUCHET C., BENHAMOU F., "The octagon abstract domain for continuous constraints", *Constraints*, vol. 19, no. 3, pp. 309–337, 2014.

[PER 09] PERRIQUET O., BARAHONA P., "Constraint-based strategy for pairwise RNA secondary structure prediction", *Proceedings of the 14th Portuguese Conference on Artificial Intelligence: Progress in Artificial Intelligence (EPIA)*, Lecture Notes in Computer Science, Springer-Verlag, vol. 5816, pp. 86–97, 2009.

[PET 11] PETIT T., RÉGIN J.-C., BELDICEANU N., "A $\Theta(n)$ bound-consistency algorithm for the increasing sum constraint", *Proceedings of the 17th International Conference on Principles and Practice of Constraint Programming (CP)*, Lecture Notes in Computer Science, Springer-Verlag, vol. 6876, pp. 721–728, 2011.

[POL 06] POLYSPACE, Institute for Radiological Protection and Nuclear Safety Verifies Nuclear Safety Software, available at http://www.mathworks.com/company/userstories/institute-for-radiological-protection-and-nuclear-safety-verifies-nuclear-safety-software.html, 2006.

[POL 10] POLYSPACE ANALYSER, available at http://www.mathworks.fr/products/polyspace, 2010.

[PON 11] PONSINI O., MICHEL C., RUEHER M., "Refining abstract interpretation-based approximations with constraint solvers", *Proceedings of the 4th International Workshop on Numerical Software Verification*, 2011.

[PRU 14] PRUD'HOMME C., FAGES J.-G., LORCA X., Choco3 Documentation, TASC, INRIA Rennes, LINA CNRS UMR 6241, COSLING S.A.S., 2014.

[PUG 98] PUGET J.-F., "A fast algorithm for the bound consistency of alldiff constraints", *Proceedings of the 15th National/10th Conference on Artificial Intelligence/Innovative Applications of Artificial Intelligence (AAAI/IAAI)*, American Association for Artificial Intelligence, pp. 359–366, 1998.

[QUI 03] QUIMPER C.-G., VAN BEEK P., LÓPEZ-ORTIZ A. *et al.*, "An Efficient Bounds Consistency Algorithm for the Global Cardinality Constraint", *Proceedings of the 9th International Conference on Principles and Practice of Constraint Programming (CP)*, Lecture Notes in Computer Science, Springer Berlin/Heidelberg, vol. 2833, pp. 600–614, 2003.

[RAM 11] RAMAMOORTHY V., SILAGHI M.C., MATSUI T. *et al.*, "The design of cryptographic S-boxes using CSPs", *Proceedings of the 17th International Conference on Principles and Practice of Constraint Programming (CP)*, Lecture Notes in Computer Science, Springer-Verlag, vol. 6876, pp. 54–68, 2011.

[RAT 94] RATZET D., "Box-splitting strategies for the interval Gauss-Seidel step in a global optimization method", *Computing*, vol. 53, pp. 337–354, 1994.

[RIV 07] RIVAL X., MAUBORGNE L., "The trace partitioning abstract domain", *ACM Transactions on Programming Languages and Systems (TOPLAS)*, vol. 29, no. 5, 2007.

[ROB 99] ROBINSON S., "Beyond 2000: further troubles lurk in the future of computing", *The New York Times*, July 1999.

[ROD 04] RODRIGUEZ-CARBONELL E., KAPUR D., "An abstract interpretation approach for automatic generation of polynomial invariants", *Proceedings of the 11th International Static Analysis Symposium (SAS)*, Lecture Notes in Computer Science, Springer, vol. 3148, pp. 280–295, 2004.

[ROS 06] ROSSI F., VAN BEEK P., WALSH T., *Handbook of Constraint Programming (Foundations of Artificial Intelligence)*, Elsevier Science Inc., New York, NY, USA, 2006.

[SCH 01] SCHULTE C., TACK G., "Implementing efficient propagation control", *Proceedings of the 3rd Workshop on Techniques for Implementing Constraint Programming Systems*, 2001.

[SCH 02] SCHULTE C., *Programming Constraint Services: High-level Programming of Standard and New Constraint Services*, Springer-Verlag, 2002.

[SCH 05] SCHULTE C., STUCKEY P.J., "When do bounds and domain propagation lead to the same search space?", *ACM Transactions on Programming Languages and Systems (TOPLAS)*, vol. 27, no. 3, pp. 388–425, 2005.

[SEL 02] SELLMANN M., "An arc-consistency algorithm for the minimum weight all different constraint", *Proceedings of the 8th International Conference on Principles and Practice of Constraint Programming (CP)*, Lecture Notes in Computer Science, Springer-Verlag, vol. 2470, pp. 744–749, 2002.

[SIM 06] SIMON A., KING A., "Widening polyhedra with landmarks", KOBAYASHI N. (ed.), *Proceedings of the 4th Asian Symposium on Programming Languages and Systems (APLAS)*, Lecture Notes in Computer Science, Springer, vol. 4279, pp. 166–182, 2006.

[SIM 10] SIMON A., CHEN L., "Simple and precise widenings for *H*-polyhedra", UEDA K. (ed.), *Proceedings of the 8th Asian Symposium on Programming Languages and Systems (APLAS)*, Lecture Notes in Computer Science, Springer, vol. 6461, pp. 139–155, 2010.

[SIM 12] SIMON B., COFFRIN C., VAN HENTENRYCK P., "Randomized adaptive vehicle decomposition for large-scale power restoration", *Proceedings of the 9th International Conference on Integration of AI and OR Techniques in Constraint Programming for Combinatorial Optimization Problems (CPAIOR)*, Lecture Notes in Computer Science, Springer-Verlag, vol. 7298, pp. 379–394, 2012.

[SOU 07] SOUYRIS J., DELMAS D., "Experimental assessment of astrée on safety-critical avionics software", *Proceedings of the 26th International Conference on Computer Safety, Reliability, and Security*, pp. 479–490, 2007.

[STØ 11] STØLEVIK M., NORDLANDER T.E., RIISE A. *et al.*, "A hybrid approach for solving real-world nurse rostering problems", *Proceedings of the 17th International Conference on Principles and Practice of Constraint Programming (CP)*, Lecture Notes in Computer Science, Springer-Verlag, vol. 6876, pp. 85–99, 2011.

[TRU 10] TRUCHET C., PELLEAU M., BENHAMOU F., "Abstract domains for constraint programming, with the example of octagons", *International Symposium on Symbolic and Numeric Algorithms for Scientific Computing*, IEEE Computer Society, pp. 72–79, 2010.

[TRU 11] TRUCHET C., ASSAYAG G. (eds.), *Constraint Programming in Music*, ISTE, London and John Wiley & Sons, New York, 2011.

[WOL 04] WOLINSKI C., KUCHCINSKI K., GOKHALE M., "A constraints programming approach to communication scheduling on SoPC architectures", *Proceedings of the ACM/SIGDA 12th International Symposium on Field Programmable Gate Arrays (FPGA)*, ACM, pp. 252–252, 2004.

# Index

**T, U**